W9-APB-198

DATE DUE

DEMCO 38-296

Genetic Engineering

Other books in the Current Controversies series:

Genetic Engineering

Lisa Yount, *Book Editor*

Daniel Leone, *President*
Bonnie Szumski, *Publisher*
Scott Barbour, *Managing Editor*

CURRENT CONTROVERSIES

GREENHAVEN PRESS
SAN DIEGO, CALIFORNIA

THOMSON
™
GALE

Detroit • New York • San Diego • San Francisco
Boston • New Haven, Conn. • Waterville, Maine
London • Munich

Cover photo: © P. Dumas/Eurelios

Library of Congress Cataloging-in-Publication Data

Genetic engineering / Lisa Yount, book editor.
    p. cm. — (Current controversies)
    Includes bibliographical references and index.
    ISBN 0-7377-1123-X (pbk. : alk. paper) —
ISBN 0-7377-1124-8 (lib. bdg. : alk. paper)
    1. Genetic engineering—Social aspects. 2. Genetic engineering—
Moral and ethical aspects. I. Yount, Lisa. II. Series.

QH442 .G44312   2002
174'.966—dc21
                                        2001051280

Copyright © 2002 by Greenhaven Press,
an imprint of The Gale Group
10911 Technology Place, San Diego, CA 92127
Printed in the U.S.A.

Contents

Chapter 1: Will Genetic Engineering Benefit Humanity?

Yes: Genetic Engineering Will Benefit Humanity

Although the knowledge behind genetic engineering shows that living things are collections of physical characteristics, this point of view does not have to decrease respect for life. While it can lead scientists to "play God," if they do so carefully, the results can be beneficial.

Gene therapy and human gene enhancement will help people live longer lives and produce healthier children. Genetically engineered crops will benefit the environment by letting farmers use less pesticides and other chemicals.

No: Genetic Engineering Will Not Benefit Humanity

Genetic engineering is based on a reductionist view of nature. It will encourage people to see other life forms, and even other human beings, as manufactured products to be controlled and changed at will.

Uncontrolled alteration of human genes threatens society and the definition of humanity, and alteration of food-crop genes threatens human health and the ecosystem. Large companies and politicians lie about these dangers.

Chapter 2: Will Genetic Engineering Benefit Food and Farming?

increase the world's food supply. Third-World farmers and governments have many good reasons for rejecting genetically engineered crops.

Chapter 3: Is Engineering Human Genes Ethical?

Yes: Engineering Human Genes Is Ethical

No: Engineering Human Genes Is Not Ethical

Chapter 4: Should Genetic Engineering Be More Closely Regulated?

Yes: Genetic Engineering Should Be More Closely Regulated

No: Genetic Engineering Need Not Be More Closely Regulated

encouraging them to share information. Patents do not block preexisting uses of plants or technologies.

Foreword

By definition, controversies are "discussions of questions in which opposing opinions clash" (Webster's Twentieth Century Dictionary Unabridged). Few would deny that controversies are a pervasive part of the human condition and exist on virtually every level of human enterprise. Controversies transpire between individuals and among groups, within nations and between nations. Controversies supply the grist necessary for progress by providing challenges and challengers to the status quo. They also create atmospheres where strife and warfare can flourish. A world without controversies would be a peaceful world; but it also would be, by and large, static and prosaic.

The Series' Purpose

The purpose of the Current Controversies series is to explore many of the social, political, and economic controversies dominating the national and international scenes today. Titles selected for inclusion in the series are highly focused and specific. For example, from the larger category of criminal justice, Current Controversies deals with specific topics such as police brutality, gun control, white collar crime, and others. The debates in Current Controversies also are presented in a useful, timeless fashion. Articles and book excerpts included in each title are selected if they contribute valuable, long-range ideas to the overall debate. And wherever possible, current information is enhanced with historical documents and other relevant materials. Thus, while individual titles are current in focus, every effort is made to ensure that they will not become quickly outdated. Books in the Current Controversies series will remain important resources for librarians, teachers, and students for many years.

In addition to keeping the titles focused and specific, great care is taken in the editorial format of each book in the series. Book introductions and chapter prefaces are offered to provide background material for readers. Chapters are organized around several key questions that are answered with diverse opinions representing all points on the political spectrum. Materials in each chapter include opinions in which authors clearly disagree as well as alternative opinions in which authors may agree on a broader issue but disagree on the possible solutions. In this way, the content of each volume in Current Controversies mirrors the mosaic of opinions encountered in society. Readers will quickly realize that there are many viable answers to these complex issues. By questioning each author's conclusions, students and casual readers can begin to develop the critical thinking skills so important to evaluating opinionated material.

Current Controversies is also ideal for controlled research. Each anthology in the series is composed of primary sources taken from a wide gamut of informational categories including periodicals, newspapers, books, United States and foreign government documents, and the publications of private and public organizations. Readers will find factual support for reports, debates, and research papers covering all areas of important issues. In addition, an annotated table of contents, an index, a book and periodical bibliography, and a list of organizations to contact are included in each book to expedite further research.

Perhaps more than ever before in history, people are confronted with diverse and contradictory information. During the Persian Gulf War, for example, the public was not only treated to minute-to-minute coverage of the war, it was also inundated with critiques of the coverage and countless analyses of the factors motivating U.S. involvement. Being able to sort through the plethora of opinions accompanying today's major issues, and to draw one's own conclusions, can be a complicated and frustrating struggle. It is the editors' hope that Current Controversies will help readers with this struggle.

Greenhaven Press anthologies primarily consist of previously published material taken from a variety of sources, including periodicals, books, scholarly journals, newspapers, government documents, and position papers from private and public organizations. These original sources are often edited for length and to ensure their accessibility for a young adult audience. The anthology editors also change the original titles of these works in order to clearly present the main thesis of each viewpoint and to explicitly indicate the opinion presented in the viewpoint. These alterations are made in consideration of both the reading and comprehension levels of a young adult audience. Every effort is made to ensure that Greenhaven Press accurately reflects the original intent of the authors included in this anthology.

Introduction

In July 2000, Larry Bohlen, a senior official of the environmentalist group Friends of the Earth, went to his local supermarket in Silver Spring, Maryland, where he purchased chips, tacos, and "all the [other] corn products I could find." He had a laboratory analyze them to determine if they contained any corn that had been genetically modified, or genetically engineered. The laboratory found that taco shells made by Kraft and sold under the brand name of Taco Bell contained traces of a type of genetically altered corn called Starlink. The Environmental Protection Agency (EPA), one of three government agencies in the United States that regulates genetically modified products, had approved Starlink for animal feed but not for human consumption. The Food and Drug Administration (FDA), which regulates the purity of foods, had not approved it either.

Friends of the Earth and another group called Genetically Engineered Food Alert, which had made a similar discovery at about the same time, told Kraft about the contaminated corn. In late September, Kraft, after doing its own tests and getting the same results, voluntarily recalled millions of its taco shells. Several other food suppliers did the same shortly afterward. Although Starlink was never shown to have harmed anyone, concern about its presence in food made headlines and had major effects on corn sales both in the United States and overseas.

Starlink corn's safety was questioned because it contained a protein called Cry9, made by a gene from a bacterium that had been transferred into the corn's genome, or collection of genes. The genetically engineered protein kills the wormlike larvae of certain insects that eat corn plants. Some groups feared that severe and even life-threatening allergic reactions might occur in humans eating corn products containing Cry9. The EPA had allowed crops that had been sprayed with this bacterium (called Bt) to be used in human food, but Cry9 had been shown to remain in the body longer than any bacterial products left by this process. It might, therefore, have a greater chance of producing an allergic reaction.

The Starlink controversy is typical of the many disputes that have arisen over genetic engineering. People have disagreed about both the safety and the ethics of altering genes or transferring them from one kind of living thing to another ever since the early 1970s, when the technology first showed promise. Supporters have claimed that genetic engineering will cure disease and decrease world hunger, but critics say it threatens both human health and the environment.

A major concern with genetic engineering has always been safety. When the procedure was in its infancy and most gene experiments were done on bacteria, the chief safety fear was that an altered microbe would escape from scientists' laboratories, causing a deadly epidemic that would resist drug therapy. Today, critics warn of more subtle dangers to health, such as the possible allergic reactions that "foreign" proteins like Cry9—substances that humans would never normally eat—might produce. They also say that genetically engineered products like Starlink threaten the environment. For instance, they point to studies suggesting that pollen from plants containing Bt genes can kill monarch butterfly caterpillars when wind blows it onto the plants that the caterpillars eat.

Supporters of genetic engineering say that these safety fears are overblown. They point to the fact that, almost 30 years after genetic engineering began, no genetically engineered living thing has ever been shown conclusively to harm health or the environment. Although some people said they felt sick after eating corn products that contained Starlink, tests showed no evidence that their bodies had actually reacted to Cry9. Far from being threats to the environment, these supporters say, Bt-containing plants like Starlink corn can protect their surroundings by limiting farmers' need to use dangerous chemical pesticides.

As with many other controversies involving genetic engineering, the safety concerns brought up in the Starlink case led to questions about the way genetic engineering and its products are regulated. "I . . . feel the [Food and Drug Administration] has failed the American public in safeguarding our food against engineered products rushed to market without adequate safety and environmental testing," Friends of the Earth's Larry Bohlen said. He and other critics say that genetically engineered foods and other products should be more thoroughly tested before they are sold and that foods containing them should be labeled so that consumers can avoid buying them if they wish.

The FDA, however, has ruled that genetically engineered foods must undergo extra testing and labeling only if they contain unusual substances such as Cry9 at the time they are eaten. Most genetically engineered crops do not fall into this category. Furthermore, Douglas Powell, a professor of plant agriculture at the University of Guelph in Ontario, Canada, and a supporter of genetically engineered crops, argues that labeling would actually reduce consumer choice by encouraging supermarkets not to carry "controversial" genetically engineered products that consumers might ultimately find desirable. Labels on genetically engineered products, he claims, would be "designed to alarm rather than inform."

To many people, the safety and regulation issues raised by debates like the one over Starlink reflect a deeper concern about the philosophy underlying genetic engineering. They see genes as the basic "stuff of life" and feel that scientists who alter genes are "playing God" and trying to reduce living things—including human beings, whose genes have also been altered—to mere manufactured products that can be modified at will. Supporters say that genetic engineering is no more disrespectful of life than any other science and that it can repair nature's mistakes, such as inherited diseases. In any case, philosopher Thomas W. Clark expressed the obvious when he wrote that since genetic engineering already has been invented and put into widespread use, "It's too late not to play God. . . . The question . . . is . . . what sorts of local gods we will or should become."

The debate over Starlink and other genetically engineered products in food is just one of the controversies surrounding genetic engineering today. Equally heated discussions rage over the usefulness of gene-altered crops in the Third World, the safety and ethics of altering human genes, and the patenting of genes, to name just a few. *Genetic Engineering: Current Controversies* focuses on these and related topics in the ongoing debate about which forms of genetic engineering, if any, should be considered safe and ethical.

Chapter 1

Will Genetic Engineering Benefit Humanity?

Chapter Preface

Much of the debate over genetic engineering concerns practical issues, such as whether gene alteration or its products will help or harm humans or the natural environment. Some commentators, however, are concerned with the more subtle question of what the power to change genes will do to human values.

Whether people realize it or not, technology has a major effect on the way they see the world. Both supporters and critics feel that genetic engineering may change the way people see other forms of life because it increases humans' power to alter and control nature. Supporters say that it could increase people's feeling of responsibility toward the environment because it increases their power to preserve the natural world or undo damage that humans have already done to it. For example, genetically engineered crops may allow farmers to use fewer chemical pesticides, and genetically engineered bacteria may help in cleaning up toxic waste.

Critics of genetic engineering, on the other hand, fear that this new power will simply intensify the view that humans are free to exploit nature at will. People will come to see the natural world not only as something to be controlled but as something to be, in a sense, manufactured to their specifications. They will forget that that world has an existence independent of, and far older than, humankind—a mere single species among the millions on the planet.

Other commentators point out that the power itself may prove to be an illusion. The ability to change genes does not necessarily include the ability to predict, let alone control, the effects of those changes. A better understanding of the complexity of genes, supporters say, could lead to a new and beneficial humility in the face of the infinitely complicated world of nature and evolution. Critics, however, believe that that humility, if it comes, may not appear soon enough, and that without it, genetic engineers are little more than children playing with matches.

At the same time genetic engineering changes values, it will be changed by them. Like any other technology, this one can be used for good or ill. The way it is used will depend on the beliefs of those who use, fund, and regulate it.

Genetic Engineering Need Not Decrease Respect for Life

by Thomas W. Clark

About the author: *Thomas W. Clark is a frequent contributor to the* Humanist, *a supporter of the philosophy of naturalism, and a research associate at Health and Addictions Research, Inc., in Boston, a nonprofit firm that conducts research on addictions, behavioral health, and criminal justice.*

The December 1999 issue of *Science* magazine reported that scientists may eventually pin down a "minimum genome": the bare bones, molecularly speaking, of what it takes to make a living organism. The biological interplay of DNA, proteins, and other subcellular components in supporting the necessary functions of life—in this case, a very simple bacterium—would be completely understood. Nothing mysterious or "protoplasmic" would remain: the very mechanism of life would stand revealed in all its complexity.

The same issue of *Science* also carried a companion piece, "The Ethical Considerations in Synthesizing a Minimal Genome," by a group of bioethicists who grapple with what they believe are the worrisome implications of such knowledge. "The attempt to model and create a minimal genome," they say, "represents the culmination of a reductionist research agenda about the meaning and significance of life that has spanned the 20th century." This agenda is far from benign, according to these ethicists, since it challenges the tradition which holds that life is valuable because it is more than "merely physical." Their worry, in essence, is that "the special status of living things and the value we ascribe to life may . . . be undermined by reductionism."

Devaluing Living Things?

This is a serious charge, one that might well tend to foster prejudice against science. If a thorough understanding of the mechanics of life necessarily devalues it, then shouldn't we pull back from the pursuit of biological knowledge? One might expect that the supposed threat of reductionism would be made clear, but in fact the authors don't sustain their indictment. Rather, their article suggests that reductionism, properly understood, poses little danger. Even with a minimum genome in hand, science simply isn't in a position to offer definitive pronouncements on the meaning or value of life.

Their worries rest on a confusion between materialism—the thesis that we essentially are physical creatures—and what might be called strong reductionism—the claim that higher level phenomena, such as human behavior, can be completely explained in terms of its underlying physical mechanisms. Now, some indeed are threatened by materialism, since being "merely" physical undercuts the traditional reassurance that the soul might outlive the body. But it's not clear that anyone should be worried about strong reductionism, since it's patently false and must be distinguished from the bread-and-butter science of analyzing biological processes, which is the work being done on the minimal genome.

The ethicists point out that "a reductionist understanding of . . . human life is not satisfying to those who believe that dimensions of the human experience cannot be explained by an exclusively physiological analysis." True enough, but does anyone really suppose that physiological analysis is even relevant to most human experience? Such strong reductionism is simply a straw man [a weak argument set up in order to be easily destroyed], not an encroaching scientific agenda.

> *"Science simply isn't in a position to offer definitive pronouncements on the meaning or value of life."*

For instance, a thorough understanding of the brain at the neural [nerve cell] level, while often necessary for tracing specific mental functions and pathologies, is simply inappropriate for dealing with the everyday psychodrama of our motives, aspirations, disappointments, and interpersonal interactions. Even though our having experiences at all may depend on our having properly wired brains, the meaning of experience derives from its social context, not its substrate [base] in physiology. In short, since analytical physical science is irrelevant to domains in which it is useless for explanatory or predictive purposes (which is to say, in much of our lives), its success cannot threaten our dignity.

Science Cannot Settle Social Issues

The ethicists also suggest that extensions of minimal genome research, by specifying the genetic definition of the human organism and its beginnings in utero [before birth], will have implications for the abortion debate. Although they don't tell us precisely what these implications are, they do conclude that "the complex metaphysical issues about the status of human beings cannot be discussed in terms of the presence or absence of a particular set of genes." Quite true, but this is yet another illustration of how physiological analysis is not about to rule our ethical intuitions. Even if we agreed on a definition of human life at the DNA level, all the contentious issues of fetal viability, birth defects, quality of life, and the sometimes conflicting interests of mother and potential newborn remain to be decided at the social level. Science simply isn't in competition with social policy debates, although it can help inform them.

But beyond abortion, the most pressing issue, the ethicists say, is

> *"It's too late not to play God."*

whether identifying minimal genomes—or perhaps even creating artificial organisms from such blueprints—"constitutes unwarranted intrusion into matters best left to nature; that is, whether work on minimal genomes constitutes 'playing God.'" How much should we intervene in the mechanics of life to suit our desires? An analytical understanding of life's mechanisms is the key to genetic engineering, both of other creatures and ourselves. If we decide we should play God, then we'll use the key; if not, we should throw it away.

What Kind of Gods Will We Be?

The authors point out that a spectrum of views exist on playing God. Many of religious persuasions reject it as arrogant hubris; others believe that it should be the no-holds-barred culmination of our capacity for self-design. They, themselves, recommend a middle path of careful biotechnological stewardship that "would move forward with caution into genomic research and with insights from valued traditions as to the proper purposes and uses of new knowledge." They also state that, "while there are reasons for caution, there is nothing in the research agenda for creating a minimal genome that is automatically prohibited by legitimate religious considerations."

If, as these ethicists conclude, there is no deep moral objection to our playing God—carefully—then a detailed analysis of life's mechanisms is simply a means to an end, not an intrinsic threat to the specialness of life or our attachment to human beings and other creatures. And it is these attachments that will shape the ends we seek and that must channel the use of biotechnology in humane, not monstrous, directions.

Were we to conclude that playing God is wrong, then advanced biology does pose a threat and we might seek to limit research into what once were the mysteries of life. Indeed, the success of science in showing that simple life forms are mechanisms—albeit astoundingly complex ones—lends power to what some feel is a deflationary materialism: we no longer need mysterious, nonphysical explanations for what life does. The sheer ability to play God, therefore, threatens those who think God is or should be a necessary hypothesis at the physical level. They would prefer science to fail, even in its proper arena, and one sure way to ensure failure is to limit biological research.

But in reality, of course, it's too late not to play God. By knowing that we have the power to know, even a decision to "let nature run its course" is yet another Godlike choice—albeit one that renounces domains of understanding and control. Such a choice would make us a god of the deists, a passive on-looker of unfolding creation, rather than an active participant in shaping our destiny.

The question, therefore, is not whether we should play God, but what sorts of local gods we will or should become. Will materialism (not the straw man of strong reductionism) demoralize us, or will we continue to find meaning in our personal and social lives even though life itself is understood to be a mechanism? The latter outcome becomes possible if we grasp that our lives' meaning need not depend on our being ethereal, as opposed to purely physical, creatures. Either way, our response to the success of science will help determine how we play the leading role in which nature has cast us on this planet.

Genetic Engineering Will Benefit Human Health and the Environment

by Jeff T. Minerd

About the author: *Jeff T. Minerd, an award winning writer, lives near Washington, D.C. He writes about science and technology for the* Futurist, *of which he is also a staff editor.*

Our understanding of DNA and our ability to manipulate it are increasing rapidly. Already we can screen human embryos for hereditary diseases such as sickle cell anemia, and we may soon treat other diseases, such as cystic fibrosis, with gene therapy. Corn and soybeans genetically altered to resist harmful pests are regularly planted in the United States. We have cloned mice and sheep. But we have barely scratched the surface of genetic knowledge. What follows is an assessment of the effects of this burgeoning field in the decades ahead.

Global sales of genetically modified crops grew from $75 million in 1995 to $1.5 billion in 1998 and are expected to reach $25 billion by 2010. The sales and profits from gene-based medicines and therapies are likely to follow a similar course once they are established. And genetic knowledge will affect many other industries: oil refining, plastics and paint production, and sanitation and waste removal, to name a few.

One analyst has gone so far as to suggest that by 2100 biotechnology will have become the dominant economic activity, similar to the status of information technology today.

A Cleaner Environment

Many genetic advancements will aid in cleaning or improving the environment. Plants engineered to "naturally" produce additives for paints and plastics will eliminate the need for environmentally unfriendly chemical processing facilities. Modified microbes may be used to clean oil spills and other toxic waste.

Crops that produce their own pesticides will reduce the need for spraying.

Some environmentalists worry about possible "genetic pollution." Perhaps the most-likely scenario is that a transgenic crop could pollinate a wild weed cousin, giving rise to "super weeds" resistant to pests and herbicides. It should be noted, however, that such a scenario is also possible with new hybrid crop strains created by the traditional method of cross-fertilization.

DNA research has revealed how closely related all life on this planet really is. Humans have many of the same genes as other forms of life, including plants, bacteria, and other animals, indicating we all evolved from a common ancestor. Once this knowledge sinks in, we may feel stronger ties of kinship with other forms of life, stimulating new waves of environmental awareness and activism.

Healthier Children

Besides ensuring that our children are born without genetic defects, we will soon be able to give them genetic enhancements: They will become taller, stronger, smarter.

By mid-twenty-first century, scientists may be able to create life from scratch in a laboratory, starting with single-celled creatures but perhaps moving to multi-cellular organisms. And once we understand the genetic mechanisms behind cancer and aging, we may find ways to put off disease and death forever and achieve virtual immortality.

"We may feel stronger ties of kinship with other forms of life, stimulating new waves of environmental awareness and activism."

Some fear that genetic advancements will lead to a state-sponsored eugenics program giving rise to a genetically superior, ruling race. A more plausible scenario, especially in Western nations that value personal freedom and choice, is that genetic enhancement of offspring will be guided by parents' personal decisions.

The wealthy will have first access to genetic enhancement, but, as with any new technology, it will become more affordable as time goes on. (Think of televisions and computers.) Eventually, the majority of people will be able to afford the technology; genetic screening and "fixing" of defective embryos may become a routine part of prenatal [before birth] care.

It should be stressed that genetic enhancement doesn't guarantee results: One's environment, experience, and effort also shape one's abilities. The genetically enhanced will have to compete with the genetically ordinary to prove themselves, giving "regular" people a chance to show they can perform as well, or better.

Longer Lives

Genetically based medicines and therapies will likely increase the already established trend toward longer life-spans and the aging of populations in developed countries. In the short term, this will probably cause conflicts between the needs of the young and the old. How, for instance, will a small group of young workers pay for the retirement of the older majority? In the longer term, if we are able to stop death itself, truly young people may become rare.

Governments will have to wrestle with complex legal issues raised by our new genetic power. Will it be legal to clone yourself in order to have a "spare" body from which to harvest organs should you need them? Such a clone, kept in a state of hibernation, might be genetically altered to have very little brain and no consciousness, so it would be human in only a very limited sense.

Privacy is another sensitive issue: Who has the right to know your genetic code? What if a couple wants to use your genes for their baby and obtains a sample from your hair or skin? Would you have the right to stop them? Do you own those genes? Courts and legislators will grapple with these issues for decades to come.

Genetic Engineering Will Decrease Respect for Life

by Ronnie Cummins

About the author: *Ronnie Cummins is national director of the Organic Consumers Association and a leader of the Campaign for Food Safety, which is dedicated to building a healthy, safe, and sustainable system of food production. He is coauthor (with Ben Lilliston) of* GE Foods: A Self-Defense Guide for Consumers.

Genetic engineering is a radical new technology, one that breaks down fundamental genetic barriers—not only between species, but also between humans, animals, and plants. By combining the genes of dissimilar and unrelated species, permanently altering their genetic codes, novel organisms are created that will pass the genetic changes onto their offspring through heredity. Scientists are now snipping, inserting, recombining, rearranging, editing, and programming genetic material. Animal genes and even human genes are being inserted into plants or animals creating unimagined transgenic life forms. For the first time in history, human beings are becoming the architects of life. Bio-engineers will be creating tens of thousands of novel organisms over the next few years. As of August 1998 no less than 37 genetically engineered foods and crops have been approved for commercialization in the US—with absolutely no pre-market safety testing or labeling required. The prospect, or rather the reality, of the Biotech Century that lies ahead is frightening.

Genetic engineering poses unprecedented ethical and social concerns, as well as serious challenges to the environment, human health, animal welfare, and the future of agriculture. The following is just a sampling of concerns:

Dangerous Technology

Genetically engineered organisms that escape or are released from the laboratory can wreak environmental havoc.

Genetically engineered "biological pollutants" have the potential to be even more destructive than chemical pollutants. Because they are alive, genetically engineered products can reproduce, migrate, and mutate. Once released, it will be virtually impossible to recall genetically engineered organisms. A report published by 100 top American scientists warned that the release of gene-spliced organisms ". . . could lead to irreversible, devastating damage to the ecology."

Genetically engineered products do not have a good track record for human safety. In 1989 and 1990, a genetically engineered brand of L-tryptophan, a common dietary supplement, killed more than 30 Americans and permanently disabled or afflicted more than 1,500 others with a potentially fatal and painful blood disorder, eosinophilia myalgia syndrome, before it was recalled by the Food and Drug Administration (FDA). The manufacturer, Showa Denko K.K., Japan's third largest chemical company, had used genetically engineered bacteria to produce the over-the-counter supplement. It is believed that the bacteria somehow became contaminated during the recombinant DNA process. There were no labels to identify the product as having been genetically engineered.

"Manufactured" Creatures

The patenting of genetically engineered foods, and widespread biotech food production, will eliminate farming as it has been practiced since the beginning. If the trend is not stopped, the patenting of transgenic plants and food-producing animals will soon lead to tenant farming in which farmers will lease their plants and animals from biotech conglomerates and pay royalties on seeds and offspring. Eventually, within the next few decades, agriculture will move off the soil and into biosynthetic industrial factories controlled by chemical and biotech companies. Never again will people know the joy of eating naturally produced fresh foods. Hundreds of millions of farmers and other workers worldwide will lose their livelihoods. The hope of creating a human, sustainable agricultural system will be destroyed.

The genetic engineering and patenting of animals reduces living beings to the status of manufactured products and will result in much suffering. In January 1994, then-U.S. Department of Agriculture (USDA) Secretary Mike Espy announced that USDA scientists had completed genome "road maps" for cattle and pigs, a precursor to ever more experimentation on live animals. In addition to the cruelty inherent in such experimentation (the mistakes are born with painful deformities, crippled, blind, and so on), these "manufactured"

"Genetic engineering poses unprecedented ethical and social concerns."

creatures have no greater value to their "creators" than mechanical inventions. Animals genetically engineered for use in laboratories, such as the infamous "Harvard mouse" which contains a human cancer-causing gene that will be passed down to all succeeding generations, were created to suffer.

A purely reductionist science, biotechnology reduces all life to bits of information (genetic code) that can be arranged and rearranged at whim. Stripped of their integrity and sacred qualities, animals that are merely objects to their "inventors" will be treated as such. Currently, more than 200 genetically engineered "freak" animals are awaiting patent approval from the federal government.

Inadequate Safety Tests

No one is regulating genetically engineered organisms adequately or properly testing them for safety. In 1986, Reagan-era policymakers stitched together a patchwork of pre-existing and only marginally appropriate statutes to ease the way for new biotechnology products. But these laws were created years ago to deal with chemicals—not the unpredictable living products of genetic engineering. To date, no suitable government apparatus has been set up to deal with this radical new class of potentially overwhelming environmental and health threats.

The FDA's policy on genetically altered foods illustrates the problem. In May 1992, then Vice President Dan Quayle, head of the Competitiveness Council, announced the US Food and Drug Administration's newly developed policy on biotech foods: genetically engineered foods will not be treated differently from naturally produced foods; they will not be safety tested; they will not carry labels stating that they have been genetically engineered, nor will the government keep track of foods that have been genetically engineered. As a result, neither the government nor consumers will know which whole or processed foods have been genetically engineered.

Vegetarians and followers of religious dietary restrictions face the prospect of unwittingly eating vegetables and fruits that contain genetic material from animals—including humans. And health risks will be discovered only by trial and error—by consumers. USDA oversight [regulation] is no better. This agency has the conflicting task of both promoting and regulating agriculture, including genetically engineered plants and animals used for food. Indeed, the USDA is a primary sponsor of biotech research on plants and animals.

By patenting the genes they discover and the living organisms they create, a small corporate elite will soon own and control the genetic heritage of the planet. Scientists who "discover" genes and ways of manipulating them can patent not only genetic engineering techniques, but also the very genes themselves. Chemical, pharmaceutical, and biotech companies such as DuPont, Upjohn, Bayer, Dow, Monsanto, Ciba-Geigy, and Rhone-Poulenc, are urgently trying to identify and patent plant, animal, and human genes in order to complete their take-over of agriculture, animal husbandry, and food processing. These are some of the same companies that once promised a carefree life through pesticides and plastics. Would you trust them with the blueprints of life?

Threats to Humanity

Genetic screening will likely lead to a loss of privacy and new levels of discrimination. Already, people are being denied health insurance on the basis of "faulty" genes. Will employers require genetic screening of their employees and deny them work on the basis of the results? Will the government have access to our personal genetic profiles? One can easily imagine new levels of discrimination being directed against those whose genetic profiles reveal them to be, for example, less intelligent or predisposed to developing certain illnesses.

Genetic engineering is already being used to "improve" the human race, a practice called eugenics. Genetic screening already allows us to identify and abort fetuses that carry genes for certain hereditary disorders. But within the next decade, scientists will likely have a complete map of the human genome to work with. Will we abort fetuses on the basis of non-life-threatening impairments such as myopia [nearsightedness], because someone is predisposed towards homosexuality, or for purely cosmetic reasons?

Researchers at the University of Pennsylvania have applied for a patent to genetically alter sperm cells in animals so traits passed down from one generation to the next can be changed; the application suggests that this can be done in humans too. Scientists are now routinely cloning sheep, mice, and soon other animals. Moving from animal eugenics to human eugenics is one small step. Everyone wants the best for their children; but where do we stop? Inadvertently, we could soon make the efforts of the Nazis to create a "superior" race seem bumbling and inefficient.

> *"Genetic screening will likely lead to a loss of privacy and new levels of discrimination."*

The US military is building an arsenal of genetically engineered biological weapons. Although the creation of biological weapons for offensive purposes has been outlawed by international treaty, the US continues to develop such weapons for defensive purposes. However, genetically engineered biological agents are identical whether they are used for offensive or defensive purposes. Areas of investigation for such weapons include: bacteria that can resist all antibiotics; extra-hardy, more virulent bacteria and viruses that live longer and kill faster; and new organisms that can defeat vaccines or natural human or plant resistances. Also being studied are the development of pathogens that can disrupt human hormonal balance enough to cause death, and the transformation of innocuous bacteria (such as are found in human intestines) into killers. Some experts believe that genetically engineered pathogens that can target specific racial groups are being developed as well.

Not all scientists are sanguine [optimistic] about genetic engineering. Among the doubters is Erwin Chargoff, the eminent biochemist who is often referred to as the father of molecular biology. He warned that all innovation does not result in "progress." Chargoff once referred to genetic engineering as "a molecular Auschwitz" and warned that the technology of genetic engineering poses a greater threat to the world than the advent of nuclear technology. "I have the feeling that science has transgressed a barrier that should have remained inviolate," he wrote in his autobiography, *Heraclitean Fire*. Noting the "awesome irreversibility" of genetic engineering experiments being planned, Chargoff warned that ". . . you cannot recall a new form of life. . . It will survive you and your children and your children's children. An irreversible attack on the biosphere is something so unheard-of, so unthinkable to previous generations, that I could only wish that mine had not been guilty of it."

Genetic Engineering Will Harm Human Health and the Environment

by Eduardo Galeano

About the author: *Eduardo Galeano is an Uruguayan journalist. His books include* Memory of Fire *and* Open Veins of Latin America.

In his novel *Brave New World,* Aldous Huxley predicted the assembly-line production of human beings. Embryos would be developed in test tubes according to their future social functions, from those created to command to those made for servitude.

Now, seventy years later, biogenetics promises us, as a sort of millennium gift, a new human race. Altering the genetic code for generations to come, science will produce beings that are intelligent, beautiful, healthy, and perhaps immortal, depending on how much money the parents have to spend.

Nobel laureate James Watson, who discovered the structure of DNA and formerly headed the Human Genome Project, preaches a despotism of science. He refuses to accept any limit on the manipulation of human reproductive cells, either for research or commerce. Without mincing words, he stated: "We have to stay away from rules and regulations."

Gregory Pence, a professor of medical ethics at the University of Alabama, is an advocate of the right of parents to pick the children they wish to have, just as "great breeders try to match a breed of dog to the needs of a family."

Lester Thurow, economist at the Massachusetts Institute of Technology and the successful theorist of success, asks who would turn down the chance to program a child with greater intelligence. "And if you don't," he argues, "your neighbors will, and your child will be the stupidest in the neighborhood."

If we are lucky, the nurseries of the future will produce superbabies like these geniuses. Today the improvement of the species doesn't require the gas chambers that Germany employed to purify its race, nor the surgery that the United

States, Sweden, and other countries use[d] to prevent low-quality human models from reproducing. The world will manufacture genetically modified people just as today it makes genetically modified (GM) foods.

Hidden Dangers

Stanley Kubrick got it right when he predicted thirty years ago in his film *2001: A Space Odyssey* that we'd be eating chemical food. The giants of the chemical industry are feeding us now. We are part of a procession of abbreviations: after DDT and PCBs, which were finally banned after it became known, years ago, that they caused more cancer than comfort, it was GMs' turn. And now GMs from the United States, Canada, and Argentina are invading the whole world, and we are all the guinea pigs in these gastrological experiments of the major laboratories.

> *"The world will manufacture genetically modified people just as today it makes genetically modified (GM) foods."*

In reality, we don't even know what we eat. Except for a few exceptions, the labels on the foods we buy don't tell us whether their ingredients have been genetically modified. Monsanto, the main provider of these products, doesn't include this detail on its labels. And the milk from cows that have been treated with its transgenic growth hormones carries no warning, though, according to studies published in *The Lancet, Science, The International Journal of Health Services,* and others, it is linked to breast and prostate cancer.

Nonetheless, the U.S. Food and Drug Administration has authorized the sale of this milk without mention of this fact on the label, because in the end hormones stimulate growth and increase production, and more production means more profit. And it's the health of the economy that comes first.

Anyway, when Monsanto is required to admit what it's selling, nothing much changes. A few years ago, the company had to pay a fine for "seventy-five inexact mentions" on the cans of its poisonous herbicide Roundup. The company was given a special bulk rate and paid a mere $3,000 per lie.

Speaking Out in Europe

The Europeans are the only ones who are defending themselves, or at least trying to. The importing of GM products is prohibited in certain cases and subject to regulation in others. Since 1998, for example, the European Union has required clear labeling on genetically modified soy products. But it is hard to put this good intention into practice: The trace of this substance is lost when combined with other ingredients. According to Greenpeace, GM soy is present in 60 percent of all the processed food sold in supermarkets around the world.

The attitude of the Europeans was shaped under the pressure of public opinion. When French farmers set fire to the silos of transgenic corn to protest the damage it does to the ecosystem, the agitator/organizer Jose Bove became a national hero, . . . who stated in his defense: "When were we, the farmers and the consumers, consulted about this?"

The government, which had arrested him, withdrew its authorization of the cultivation of biotech corn.

Of course, the Europeans have other reasons to distrust technocrats' maneuvers on their dinner tables. They are still shaken by their recent experience with mad cows. For the thousands of years that cows lived off grass and grain, their behavior was impeccable and they accepted their fate with resignation. Then our insane current system forced them into cannibalism. Cows ate cows and grew fatter, rendered humanity more meat and milk, won the applause of the markets and encomiums from their owners—and went stark raving mad. People made a lot of jokes about this—until they started to die from it. One, then ten, then twenty, a hundred. . . .

In 1996, the British Ministry of Agriculture informed the population that animal feed made from animal blood, fat, and gelatin was safe for cattle and not harmful to human health. Bon appetit!

Chapter 2

Will Genetic Engineering Benefit Food and Farming?

The Perils and Promise of Genetically Engineered Crops: An Overview

by *Time*

About the author: Time *is a national weekly news magazine.*

At first, the grains of rice that Ingo Potrykus sifted through his fingers did not seem at all special, but that was because they were still encased in their dark, crinkly husks. Once those drab coverings were stripped away and the interiors polished to a glossy sheen, Potrykus and his colleagues would behold the seeds' golden secret. At their core, these grains were not pearly white, as ordinary rice is, but a very pale yellow—courtesy of beta-carotene, the nutrient that serves as a building block for vitamin A.

Potrykus was elated. For more than a decade he had dreamed of creating such a rice: a golden rice that would improve the lives of millions of the poorest people in the world. He'd visualized peasant farmers wading into paddies to set out the tender seedlings and winnowing the grain at harvest time in handwoven baskets. He'd pictured small children consuming the golden gruel their mothers would make, knowing that it would sharpen their eyesight and strengthen their resistance to infectious diseases.

And he saw his rice as the first modest start of a new green revolution, in which ancient food crops would acquire all manner of useful properties: bananas that wouldn't rot on the way to market; corn that could supply its own fertilizer; wheat that could thrive in drought-ridden soil.

But imagining a golden rice, Potrykus soon found, was one thing and bringing one into existence quite another. Year after year, he and his colleagues ran into one unexpected obstacle after another, beginning with the finicky growing habits of the rice they transplanted to a greenhouse near the foothills of the Swiss Alps. When success finally came, in the spring of 1999, Potrykus was 65 and about to retire as a full professor at the Swiss Federal Institute of Technology in Zurich. At that point, he tackled an even more formidable challenge.

Having created golden rice, Potrykus wanted to make sure it reached those for whom it was intended: malnourished children of the developing world. And that, he knew, was not likely to be easy. Why? Because in addition to a full complement of genes from *Oryza sativa*—the Latin name for the most commonly consumed species of rice—the golden grains also contained snippets of DNA borrowed from bacteria and daffodils. It was what some would call Frankenfood, a product of genetic engineering. As such, it was entangled in a web of hopes and fears and political baggage, not to mention a fistful of ironclad patents.

Ever since Potrykus and his chief collaborator, Peter Beyer of the University of Freiburg in Germany, announced their achievement [in mid-1999,] their golden grain has illuminated an increasingly polarized public debate. At issue is the question of what genetically engineered crops represent. Are they, as their proponents argue, a technological leap forward that will bestow incalculable benefits on the world and its people? Or do they represent a perilous step down a slippery slope that will lead to ecological and agricultural ruin? Is genetic engineering just a more efficient way to do the business of conventional cross-breeding? Or does the ability to mix the genes of any species—even plants and animals—give man more power than he should have?

The debate erupted the moment genetically engineered crops made their commercial debut in the mid-1990s, and it has escalated ever since. First to launch major protests against biotechnology were European environmentalists and consumer-advocacy groups. They were soon followed by their U.S. counterparts, who made a big splash at the World Trade Organization meeting in Seattle [in December 1999] and [in mid-2000] launched an offensive designed to target one company after another. Charges that transgenic crops pose grave dangers will be raised in petitions, editorials, mass mailings and protest marches. As a result, golden rice, despite its humanitarian intent, will probably be subjected to the same kind of hostile scrutiny that has already led to curbs on the commercialization of these crops in Britain, Germany, Switzerland and Brazil.

The hostility is understandable. Most of the genetically engineered crops introduced so far represent minor variations on the same two themes: resistance to insect pests and to herbicides used to control the growth of weeds. And they are often marketed by large, multinational corporations that produce and sell the very agricultural chemicals farmers are spraying on

> *"[Potrykus] saw his rice as the first modest start of a new green revolution."*

their fields. So while many farmers have embraced such crops as Monsanto's Roundup Ready soybeans, with their genetically engineered resistance to Monsanto's Roundup-brand herbicide, that let them spray weed killer without harming crops, consumers have come to regard such things with mounting suspicion.

Why resort to a strange new technology that might harm the biosphere, they ask, when the benefits of doing so seem small?

Indeed, the benefits have seemed small—until golden rice came along to suggest otherwise. Golden rice is clearly not the moral equivalent of Roundup Ready beans. Quite the contrary, it is an example—the first compelling example—of a genetically engineered crop that may benefit not just the farmers who grow it but also the consumers who eat it. In this case, the consumers include at least a million children who die every year because they are weakened by vitamin-A deficiency and an additional 350,000 who go blind.

No wonder the biotech industry sees golden rice as a powerful ally in its struggle to win public acceptance. No wonder its critics see it as a cynical ploy. And no wonder so many of those concerned about the twin evils of poverty and hunger look at golden rice and see reflected in it their own passionate conviction that genetically engineered crops can be made to serve the greater public good—that in fact such crops have a critical role to play in feeding a world that is about to add to its present population of 6 billion. As former President Jimmy Carter put it, "Responsible biotechnology is not the enemy; starvation is."

Indeed, by the year 2020, the demand for grain, both for human consumption and for animal feed, is projected to go up by nearly half, while the amount of arable land available to

> *"Why resort to a strange new technology that might harm the biosphere, . . . when the benefits of doing so seem small?"*

satisfy that demand will not only grow much more slowly but also, in some areas, will probably dwindle. Add to that the need to conserve overstressed water resources and reduce the use of polluting chemicals, and the enormity of the challenge becomes apparent. In order to meet it, believes Gordon Conway, the agricultural ecologist who heads the Rockefeller Foundation, 21st century farmers will have to draw on every arrow in their agricultural quiver, including genetic engineering. And contrary to public perception, he says, those who have the least to lose and the most to gain are not well-fed Americans and Europeans but the hollow-bellied citizens of the developing world.

Going for the Gold

It was in the late 1980s, after he became a full professor of plant science at the Swiss Federal Institute of Technology, that Ingo Potrykus started to think about using genetic engineering to improve the nutritional qualities of rice. He knew that of some 3 billion people who depend on rice as their major staple, around 10% risk some degree of vitamin-A deficiency and the health problems that result. The reason, some alleged, was an overreliance on rice ushered in by the green revolution. Whatever its cause, the result was distressing: these people were so poor that they ate a few bowls of rice a day and almost nothing more.

The problem interested Potrykus for a number of reasons. For starters, he was attracted by the scientific challenge of transferring not just a single gene, as many had already done, but a group of genes that represented a key part of a biochemical pathway. He was also motivated by complex emotions, among them empathy. Potrykus knew more than most what it meant not to have enough to eat. As a child growing up in war-ravaged Germany, he and his brothers were often so desperately hungry that they ate what they could steal.

Around 1990, Potrykus hooked up with Gary Toenniessen, director of food security for the Rockefeller Foundation. Toenniessen had identified the lack of beta-carotene in polished rice grains as an appropriate target for gene scientists like Potrykus to tackle because it lay beyond the ability of traditional plant breeding to address. For while rice, like other green plants, contains light-trapping beta-carotene in its external tissues, no plant in the entire *Oryza* genus—as far as anyone knew—produced beta-carotene in its endosperm (the starchy interior part of the rice grain that is all most people eat).

It was at a Rockefeller-sponsored meeting that Potrykus met the University of Freiburg's Peter Beyer, an expert on the beta-carotene pathway in daffodils. By combining their expertise, the two scientists figured, they might be able to remedy this unfortunate oversight in nature. So in 1993, with some $100,000 in seed money from the Rockefeller Foundation, Potrykus and Beyer launched what turned into a seven-year, $2.6 million project, backed also by the Swiss government and the European Union. "I was in a privileged situation," reflects Potrykus, "because I was able to operate without industrial support. Only in that situation can you think of giving away your work free."

That indeed is what Potrykus announced he and Beyer planned to do. The two scientists soon discovered, however, that giving away golden rice was not going to be as easy as they thought. The genes they transferred and the bacteria they used to transfer those genes were all encumbered by patents and proprietary rights. [In the spring of 2000,] the two scientists struck a deal with AstraZeneca, which is based in London and holds an exclusive license to one of the genes Potrykus

> *"Cornell University entomologist John Losey . . . dusted Bt corn pollen on plants populated by monarch-butterfly caterpillars. Many of the caterpillars died."*

and Beyer used to create golden rice. In exchange for commercial marketing rights in the U.S. and other affluent markets, AstraZeneca agreed to lend its financial muscle and legal expertise to the cause of putting the seeds into the hands of poor farmers at no charge.

No sooner had the deal been made than the critics of agricultural biotechnology erupted. "A rip-off of the public trust," grumbled the Rural Advancement Foundation International, an advocacy group based in Winnipeg, Canada. "Asian farmers get (unproved) genetically modified rice, and AstraZeneca gets the 'gold.'" Potrykus was dismayed by such negative reaction. "It would be irresponsible," he exclaimed, "not to say immoral, not to use biotechnology to try to solve this problem!" But such expressions of good intentions would not be enough to allay his opponents' fears.

Weighing the Perils

Beneath the hyperbolic talk of Frankenfoods and Superweeds, even proponents of agricultural biotechnology agree, lie a number of real concerns. To begin with, all foods, including the transgenic foods created through genetic engineering, are potential sources of allergens. That's because the transferred genes contain instructions for making proteins, and not all proteins are equal. Some—those in peanuts, for example—are well known for causing allergic reactions. To many, the possibility that golden rice might cause such a problem seems farfetched, but it nonetheless needs to be considered.

Then there is the problem of "genetic pollution," as opponents of biotechnology term it. Pollen grains from such wind-pollinated plants as corn and canola, for instance, are carried far and wide. To farmers, this mainly poses a nuisance. Transgenic canola grown in one field, for example, can very easily pollinate nontransgenic plants grown in the next. Indeed this is the reason behind the furor that recently erupted in Europe when it was discovered that canola seeds from Canada—unwittingly planted by farmers in England, France, Germany and Sweden—contained transgenic contaminants.

The continuing flap over Bt corn and cotton—now grown not only in the U.S. but also in Argentina and China—has provided more fodder for debate. Bt stands for a common soil bacteria, *Bacillus thuringiensis*, different strains of which produce toxins that target specific insects. By transferring to corn and cotton the bacterial gene responsible for making this toxin, Monsanto and other companies have produced crops that are resistant to the European corn borer and the cotton bollworm. An immediate concern, raised by a number of ecologists, is whether or not widespread planting of these crops will spur the development of resistance to Bt among crop pests. That would be unfortunate, they point out, because Bt is a safe and effective natural insecticide that is popular with organic farmers.

Even more worrisome are ecological concerns. In 1999 Cornell University entomologist John Losey performed a provocative, "seat-of-the-pants" laboratory experiment. He dusted Bt corn pollen on plants populated by monarch-butterfly caterpillars. Many of the caterpillars died. Could what happened in Losey's laboratory happen in cornfields across the Midwest? Were these lovely butterflies, already under pressure owing to human encroachment on their Mexican wintering grounds, about to face a new threat from high-tech farmers in the north?

The upshot: despite studies pro and con—and countless save-the-monarch protests acted out by children dressed in butterfly costumes—a conclusive answer to this question has yet to come. Losey himself is not yet convinced that Bt corn poses a grave danger to North America's monarch-butterfly population, but he does think the issue deserves attention. And others agree. "I'm not anti biotechnology per se," says biologist Rebecca Goldberg, a senior scientist with the Environmental Defense Fund, "but I would like to have a tougher regulatory regime. These crops should be subject to more careful screening before they are released."

Are there more potential pitfalls? There are. Among other things, there is the possibility that as transgenes in pollen drift, they will fertilize wild plants, and weeds will emerge that are hardier and even more difficult to control. No one knows how common the exchange of genes between domestic plants and their wild relatives really is, but Margaret Mellon, director of the Union of Concerned Scientists' agriculture and biotechnology program, is certainly not alone in thinking that it's high time we find out. Says she: "People should be responding to these concerns with experiments, not assurances."

And that is beginning to happen, although—contrary to expectations—the reports coming in are not necessarily that scary. [Since 1997,] University of Arizona entomologist Bruce Tabashnik has been monitoring fields of Bt cotton that farmers have planted in his state. And in this instance at least, he says, "the environmental risks seem minimal, and the benefits seem great." First of all, cotton is self-pollinated rather than wind-pollinated, so that the spread of the Bt gene is of less concern. And because the Bt gene is so effective, he notes, Arizona farmers have reduced their use of chemical insecticides 75%. So far, the pink bollworm population has not rebounded, indicating that the feared resistance to Bt has not yet developed.

Assessing the Promise

Are the critics of agricultural biotechnology right? Is biotech's promise nothing more than overblown corporate hype? The papaya growers in Hawaii's Puna district clamor to disagree. In 1992 a wildfire epidemic of papaya ringspot virus threatened to destroy the state's papaya industry; by 1994, nearly half the state's papaya acreage had been infected, their owners forced to seek outside employment. But then help arrived, in the form of a virus-resistant transgenic papaya developed by Cornell University plant pathologist Dennis Gonsalves.

In 1995 a team of scientists set up a field trial of two transgenic lines—UH SunUP and UH Rainbow—and by 1996, the verdict had been rendered. As everyone could see, the nontransgenic plants in the field trial were a stunted mess, and the transgenic plants were healthy. In 1998, after negotiations with four patent holders, the papaya growers switched en masse to the transgenic seeds and reclaimed their orchards. "Consumer acceptance has been great," reports Rusty Perry, who runs a papaya farm near Puna. "We've found that customers are more concerned with how the fruits look and taste than with whether they are transgenic or not."

Viral diseases, along with insect infestations, are a major cause of crop loss in Africa, observes Kenyan plant scientist Florence Wambugu. African sweet-potato fields, for example, yield only 2.4 tons per acre, vs. more than double that in the rest of the world. Soon Wambugu hopes to start raising those yields by introducing a transgenic sweet potato that is resistant to the feathery mottle virus. There really is no other option, explains Wambugu, who currently directs the International Service for the Acquisition of Agri-biotech Applications in Nairobi.

"In [Florence] Wambugu's view, there are more benefits to be derived from agricultural biotechnology in Africa than practically anywhere else on the planet."

"You can't control the virus in the field, and you can't breed in resistance through conventional means."

To Wambugu, the flap in the U.S. and Europe over genetically engineered crops seems almost ludicrous. In Africa, she notes, nearly half the fruit and vegetable harvest is lost because it rots on the way to market. "If we had a transgenic banana that ripened more slowly," she says, "we could have 40% more bananas than now." Wambugu also dreams of getting access to herbicide-resistant crops. Says she: "We could liberate so many people if our crops were resistant to herbicides that we could then spray on the surrounding weeds. Weeding enslaves Africans; it keeps children from school."

In Wambugu's view, there are more benefits to be derived from agricultural biotechnology in Africa than practically anywhere else on the planet—and this may be so. Among the genetic-engineering projects funded by the Rockefeller Foundation is one aimed at controlling striga, a weed that parasitizes the roots of African corn plants. At present there is little farmers can do about striga infestation, so tightly intertwined are the weed's roots with the roots of the corn plants it targets. But scientists have come to understand the source of the problem: corn roots exude chemicals that attract striga. So it may prove possible to identify the genes that are responsible and turn them off.

The widespread perception that agricultural biotechnology is intrinsically inimical to the environment perplexes the Rockefeller Foundation's Conway, who views genetic engineering as an important tool for achieving what he has termed a "doubly green revolution." If the technology can marshal a plant's natural defenses against weeds and viruses, if it can induce crops to flourish with minimal application of chemical fertilizers, if it can make dryland agriculture more productive without straining local water supplies, then what's wrong with it?

Of course, these particular breakthroughs have not happened yet. But as the genomes of major crops are ever more finely mapped, and as the tools for transferring genes become ever more precise, the possibility for tinkering with complex biochemical pathways can be expected to expand rapidly. As Potrykus sees it, there is no question that agricultural biotechnology can be harnessed for the good of humankind. The only question is whether there is the collective will to do so. And the answer may well emerge as the people of the world weigh the future of golden rice.

Genetically Engineered Foods Are Safe to Grow and Eat

by the Council for Biotechnology Information

About the author: *The Council for Biotechnology Information was founded by leading biotechnology companies to provide information about the benefits of this technology.*

Q: What are the most common biotech crops? What crops are in development?

A: According to U.S. Department of Agriculture (USDA) estimates, about 55 percent of soybeans, 60 percent of cotton and 36 percent of corn grown in the United States in 1999 were derived through biotechnology. More than 60 million acres of biotech crops have been planted in the United States. Canola, corn, soybeans, and potatoes are examples of biotech crops widely grown in Canada. In 1999, more than half of the canola acreage and one third of the corn acreage was in varieties or hybrids developed using biotechnology. Researchers continue to work on a variety of new biotech crops. For example, new strains of rice and other subsistence crops would help combat vitamin A deficiency—a leading cause of blindness in the developing world—by delivering higher doses of beta-carotene.

Can foods derived through biotechnology be more nutritious than conventional foods?

Now and in the near future biotechnology products provide potential food quality improvements. Some biotech foods may help to prevent heart disease and cancer by delivering more of vitamins C and E. Research is under way on "golden rice," which would combat Vitamin A deficiency in developing nations by delivering more beta-carotene and iron. Other biotech foods under development, such as a potato that absorbs less oil, may help to prevent heart disease by cutting back on fatty acids. Biotechnology could improve nutrition in other ways, such as by producing allergy-free peanuts and rice. Researchers are even working on a banana that could deliver vaccines against Hepatitis B [a liver disease caused by a virus] and other deadly diseases.

Biotech Crops Are Regulated

How are foods derived through biotechnology regulated in the United States?
Biotech foods are extensively researched and reviewed. In the United States, three government agencies—the Food and Drug Administration (FDA), Department of Agriculture (USDA), and Environmental Protection Agency (EPA) as well as many individual state governments—work together to ensure that crops produced through biotechnology are safe.

How are foods derived through biotechnology regulated in Canada?
The Canadian government strictly regulates biotechnology foods to ensure their dietary and environmental safety. Biotechnology seeds and foods, like other products, are carefully regulated by the Canadian Food Inspection Agency, Health Canada and Environment Canada. International organizations like the United Nations, the World Health Organization and the World Trade Organization provide additional protection. Current law requires biotech foods to be labeled if their composition or nutritional content is significantly different from their conventional counterparts or if they pose any health risk. In addition, the Canadian federal government is also working with food producers and others to craft voluntary labeling standards for biotechnology foods.

If biotech products are safe, why are biotechnology companies opposed to labeling them?
We are supportive of efforts to ensure that consumers have the information they need to make sound food decisions. The question of consumer product labeling is best addressed by the food industry working in cooperation with regulatory agencies. In the U.S. and Canada, this cooperative effort has resulted in a science-based system that requires labeling if the food differs in safety, composition, or nutritional quality compared to conventional food. No products developed using biotechnology that are currently on the market fall into this category.

I'm concerned that biotech foods could contain genes from foods to which I'm allergic. How can I be sure they don't?

> *"Biotechnology products provide potential food quality improvements."*

Internationally, the evaluation of allergenic potential is an integral part of a regulatory approval process. Labeling is mandatory if a known allergen is transferred to a food source not normally associated with that allergen. Again, no products developed using biotechnology that are currently on the market fall into this category.

What is food biotechnology?
Food biotechnology covers such diverse activities as the use of yeast in brewing or bread-making to advanced plant breeding techniques. New developments

41

in biotechnology allow us to identify and transfer the specific gene that creates a desired trait in a plant, and offer a more precise way to produce plants with certain beneficial characteristics—such as greater nutrition.

Biotech Crops Don't Harm the Environment

How do foods derived through biotechnology protect the environment?

Discoveries in biotechnology allow some crops to have their own protection against insects and disease and, therefore, can be grown using less crop protection chemicals. For example, cotton and corn now can resist some destructive insects on their own. This allows farmers to choose the best combination of tools to control harmful pests and diseases. Biotechnology can provide opportunities to decrease soil erosion and greenhouse gas emissions through farming practices that protect the environment. Some of these new crops require less tilling, helping to preserve precious topsoil, use less fuel and reduce farm run-off into streams and rivers.

Are other plants and wildlife affected by growing crops developed using biotechnology?

Scientists at the Arable Crops Research Institute, an independent, publicly-funded body in the U.K. [United Kingdom, or Britain], have found that by using biotechnology innovations to grow disease-resistant crops, farmers will be able to grow crops using fewer chemicals to protect crops. This gives growers another tool in protecting plants from harmful pests and disease. Some biotech crops may even be friendlier to other plants and wildlife.

> *"Thanks to . . . developments in food biotechnology, we'll be able to grow more food and better food on land already being farmed."*

For example, cotton has been developed that has the ability to protect itself from certain insects. As a result, insects that pose no threat to the cotton crop are more able to thrive, while the insects that destroy the crop are controlled.

Are Monarch butterflies affected by corn derived through biotechnology?

A study performed at Cornell University examined what happened when Monarch butterfly larvae [caterpillars] were fed milkweed with large amounts of pollen from Bt corn. As stated by Dr. John Losey, the lead researcher of the study, ". . . our study was conducted in the laboratory . . . it would be inappropriate to draw any conclusions about the risk to monarch populations in the field based solely on these initial results." In response to this study, leading researchers, entomologists and weed scientists headed to cornfields in the summer of 1999 to assess the risk of Monarch butterfly exposure to Bt corn pollen under natural conditions. The studies show that the concentration of Bt pollen adhering to milkweeds within just a few meters of cornfields is typically too low to cause mortality of even small Monarch caterpillars that might be present during pollen shed. These latest findings provide reassurance for the safety of the Monarch butterfly in a real world situation.

Didn't one study raise questions about the safety of potatoes derived through biotechnology?

A preliminary study performed at the Rowett Research Institute in Scotland in early 1999 reported that rats developed intestinal problems after being fed raw biotech potatoes. This study was highly criticized by independent experts including the Rowett Institute itself, which discredited the study entirely after performing an audit of the research. The study was however later accepted and published in *The Lancet,* one of Britain's leading medical journals, although the same journal commented that the study proved no link between biotech potatoes and intestinal problems in the rats. As one of *The Lancet*'s independent experts said, all the study proved was that "raw potatoes are not very nutritious." Professor Martin Chrispeels of the University of California at San Diego, said of the study: "This isn't science."

I've read that biotechnology crops that resist pests, weeds and disease could lead to the evolution of "super weeds." Is this true?

Farmers have widely adopted crops improved through biotechnology because they provide them with another tool to help control insects and weeds that can damage crops or decrease productivity. Crops currently in the marketplace have undergone thorough testing in the field and have passed strict environmental regulation criteria. It is also worth noting that resistance to products used to control pests is an issue with which farmers have been dealing long before the introduction of crops developed using biotechnology. That's why the principles of Integrated Crop Management (ICM) are being increasingly adopted by farmers worldwide. ICM is designed to prevent the weeds and insects from becoming immune to control. Thanks to biotechnology, growers have another way to protect crops by enabling plants to protect themselves from harmful pests.

How can biotech crops be a part of the solution to feeding a growing population?

According to statistics from the Population Division of the United Nations Department of Economic and Social Affairs, the world population will likely increase to approximately nine billion by 2050. The fact is that with more people, we will need to provide more food. At the same time, there is little remaining land for farming, without destroying valuable rainforest and wetland habitats. World hunger is a complex issue that biotechnology can play a part in helping. Thanks to continued improvements in agriculture and food production, and to developments in food biotechnology, we'll be able to grow more food and better food on land already being farmed.

Poor Farmers Need Genetically Engineered Crops

by Hassan Adamu

About the author: *Hassan Adamu is Nigeria's minister of agricultural and rural development.*

It is possible to kill someone with kindness, literally. That could be the result of the well-meaning but extremely misguided attempts by European and North American groups that are advising Africans to be wary of agricultural biotechnology. They claim to have the environment and public health at the core of their opposition, but scientific evidence disproves their claims that enhanced crops are anything but safe. If we take their alarmist warnings to heart, millions of Africans will suffer and possibly die.

Agricultural biotechnology, whereby seeds are enhanced to instill herbicide tolerance or provide resistance to insects and disease, holds great promise for Africa and other areas of the world where circumstances such as poverty and poor growing conditions make farming difficult. Fertilizer, herbicides, pesticides, machinery, fuel and other tools that richer nations take for granted as part of their farming regimen [procedure] are luxuries in poorer countries.

Moreover, the soil in tropical climates, or in areas with inhospitable weather, cannot be farmed successfully in the more traditional ways. These circumstances demand unique agricultural solutions, and many have been made available through the advances of biotechnology.

To deny desperate, hungry people the means to control their futures by presuming to know what is best for them is not only paternalistic [condescending, treating such people like children] but morally wrong. Certainly, those with fertile lands and an abundance of food have every right to decide how they would like to grow their crops and process their foods. Organic farming, sophisticated methods of distributing food and other approaches are well and good for those who can afford to experiment. Starving people do not have this luxury. They want food and nourishment, not lectures, and we certainly won't allow ourselves to be intimidated by eco-terrorists who destroy test crops and disrupt scientific meetings that strive to reveal the facts.

It is wrong and dangerous for a privileged people to presume that they know what is best for everyone. And when this happens, it cannot come as a shock that those who are imposed upon often see this attitude as colonialist [like that taken by countries that control other countries].

Africa Needs Agricultural Biotechnology

Millions of Africans—far too many of them children—are suffering from malnutrition and hunger. Agricultural biotechnology offers a way to stop the suffering. As Florence Wambugu, one of Africa's leading plant geneticists, said recently, "In Africa, GM [genetically modified] food could almost literally weed out poverty."

With regard to agricultural biotechnology, Africans are not asking for others to come in and grow our food. We are not asking for others to provide the financial means to establish this system in our countries. We want to come to the table as stakeholders. We know the conditions of our fields. We know the threats, the insects and diseases. We can work as partners to develop the seeds that could build peoples and nations.

> *"It is wrong and dangerous for a privileged people to presume that they know what is best for everyone."*

We do not want to be denied this technology because of a misguided notion that we don't understand the dangers or the future consequences. We understand. We understand that this system must continue to undergo study and careful use. We also understand that agricultural biotechnology has been deemed safe and nutritious by a host of nationally and internationally respected organizations such as the National Research Council, Nuffield Council on Bioethics, World Health Organization, Food and Agriculture Organization of the United Nations, Organization for Economic Cooperation and Development, the American Medical Association and the American Dietetic Association.

We will proceed carefully and thoughtfully, but we want to have the opportunity to save the lives of millions of people and change the course of history in many nations. That is our right, and we should not be denied by those with a mistaken idea that they know best how everyone should live or that they have the right to impose their values on us.

The harsh reality is that, without the help of agricultural biotechnology, many will not live.

Genetic Engineering of Animals Benefits Human Health

by Gunjan Sinha

About the author: *Gunjan Sinha is an associate editor and columnist for Popular Science. He won the Ray Bruner Science Writing Award from the American Public Health Association in November 2000.*

Forget Dolly. New cloning techniques produce cows with life-giving medicines in their milk.

Just beyond the greasy burger joints lining the main street of Worcester, Massachusetts, is a laboratory where a small group of scientists is tinkering with a technology that might forever transform the way medicines are made. The scientists at this lab are creating a new kind of drug factory, one without a single piece of machinery.

It will be made from skin, bones, and udder: a very ordinary framework for a genetically engineered cow that secretes medicine for humans in its milk.

Living Drug "Factories"

Steven Stice hovers over a microscope, eyeing a single bovine cell that he hopes will grow into a medicine-making cow. If this experiment is successful, the cow will join growing herds of sheep, goats, and pigs at the forefront of a revolution in biology. A combination of two new technologies—gene insertion and cloning—is making this revolution possible. It's all happening so quickly that the first drugs made from animal milk were predicted ready for the market by 1999.

Advanced Cell Technology

Stice and a group of fellow scientists formed Worcester-based Advanced Cell Technology (ACT) in 1994. They knew that many human diseases are caused by defective proteins, and that the blueprints for these proteins are defective genes. The scientists reasoned that if they could insert healthy human genes into a cow, and then clone that cow, they might be able to create an entire herd of animals capable of producing healthy human proteins in their milk. These proteins could then be extracted from the milk and packaged as medicines. ACT scientists chose cows because the animals pump out enormous quantities of milk and would consequently produce large amounts of medicines.

ACT's first target drug is human serum-albumin, a protein component of blood that, among other functions, gives blood enough pressure for the heart to pump it efficiently through the body. Doctors inject the protein into patients undergoing various types of surgeries to maintain their blood pressure. The annual worldwide demand for serum-albumin is between 400 and 500 metric tons, according to Stice.

Currently, serum-albumin is extracted from human blood. But with mounting concerns over infectious human pathogens such as the HIV virus, human blood has become a suspect source. What's more, rigid screening procedures have shrunk the blood donor pool so dramatically that the protein is becoming scarce, which is driving up the cost. Stice estimates that a herd of between 2,000 and 3,000 cows could produce enough protein to satisfy the current demand at a reasonable cost, without the risk of transmitting infectious agents.

Producing medicines in animal milk holds so much promise that even non-profit agencies like the American Red Cross (ARC) are interested. In 1998 ARC teamed up with the Dutch drug company Pharming to produce human fibrinogen—a blood protein responsible for blood clotting—in cow milk. ARC is working with the Virginia Polytechnic Institute in Blacksburg to incorporate fibrinogen into a sophisticated new bandage that stops bleeding almost instantly. "We're talking about a bandage that could revolutionize emergency medical care," says William Drohan, head of ARC's Plasma Derivatives Laboratory in Washington, D.C. "We can also formulate the fibrinogen sealant into an expandable foam or powder that could be sprayed on deep gouges to plug severe bleeding," he adds.

Fibrinogen is currently extracted from human blood, but there just aren't enough donors to supply the protein in the amounts necessary to make the sealant. The only way of producing huge quantities is to use animals that have been genetically engineered to carry the human gene responsible for the production of fibrinogen.

"[Scientists] might be able to create an entire herd of animals capable of producing healthy human proteins in their milk."

Milk isn't the only vehicle for producing medicine. James Petitte at North Carolina State University in Raleigh is trying to insert human genes into chickens, prompting them to make human proteins in their eggs. And Bob Wall of the Agricultural Research Service in Beltsville, Maryland, is popping human genes into mice, and hoping that the mice will produce human proteins in their urine.

The ability to mix and match genes from different species—known as transgenics—has been around since the 1970s. Mice, goats, and sheep that carry genes from other species, including humans, are almost commonplace. Scientists make these animals by injecting human DNA into animal embryos, a process that involves a lot of trial and error.

Combining transgenic technology with cloning, however, speeds up the process and cuts costs by eliminating a lot of guesswork. To create a herd of transgenic cloned calves that produce serum-albumin, for example, the scientists at ACT splice human genes into fetal cow cells and then fuse the nuclei of these cells with eggs that are implanted in surrogate mothers.

> *"Pigs have proven good donors [for potential human organ transplants] because the sizes and shapes of their organs match those of humans."*

Cloning also gives scientists better control over the transgenic technique by allowing them to not only add genes, but also take them away. That could make it possible to engineer animals to produce other medically useful products—such as organs and tissues for transplantation into humans. In fact, Stice is trying to perfect a technique to genetically modify and clone pigs for precisely that purpose.

Pigs have proven good donors because the sizes and shapes of their organs match those of humans. But because porcine tissue is so different from human tissue, it immediately turns black when transplanted: Human immune cells, recognizing the pig tissue as foreign, choke its blood supply. By turning off the genes responsible for this problem, doctors might be able to transplant pig organs directly into humans without giving patients toxic immune-suppressing drugs.

Novartis Pharmaceuticals Corp, of East Hanover, New Jersey, has already had some success in transplanting genetically engineered pig hearts and kidneys into monkeys. In preliminary studies, the monkeys have been able to hold onto the transplanted organs for as long as 70 days.

Stice himself has been cloning animals since the late 1980s. "The biggest misconception about cloning is that everything started with Dolly," he says. In fact, the basic cloning techniques used today were developed in the late 1980s. At the time, the goal was to clone prized animals for agricultural purposes. Scientists hoped, for example, to clone cows with the highest milk yields, the most tender meat, and the best flavor. But most of these efforts flopped after experiments showed cloning large animals to be more difficult than anticipated. The physiology and embryology of each species varies tremendously, and the payoff was too low to justify the expense.

Combining cloning with transgenics to produce medicines makes better economic sense, because the end product, a drug, is worth a lot more than a prime cut of meat. In fact, Dolly herself was the result of an effort to create a sheep that would secrete medicine in its milk.

Today, a number of biotechnology companies around the world are pursuing this idea, using sheep and other animals in attempts to produce a variety of medicines.

Although Dolly's birth proved that an animal could be cloned from an adult cell, making copies of adult animals is not the focus of scientists trying to produce medicine in animal milk. "Why use an adult cell when we have much greater success in making cloned transgenic animals with younger fetal or embryonic cells?" Stice asks.

But that doesn't mean scientists aren't investigating the cloning of adult cells. There are many unanswered questions about how cloning works, and the technology has other applications such as cloning endangered species, growing human tissues for transplants, and generating cloned lab animals for research. Making medicine in animal milk is simply the first application of cloning. Although some of the future applications may be ethically troubling, their potential benefits are too tantalizing for scientists not to forge ahead.

Genetically Engineered Crops Threaten Farmers, Health, and the Environment

by David Ehrenfeld

About the author: *David Ehrenfeld is professor of biology at the Department of Ecology, Evolution, and Natural Resources at Rutgers University in New Brunswick, New Jersey. He writes frequently on conservation biology and is a founding editor of* Conservation Biology *magazine.*

The modern history of agriculture has two faces. The first, a happy face, is turned toward nonfarmers who live in the developed world. It speaks brightly of technological miracles, such as the "Green Revolution" and, more recently, genetic engineering, that have resulted in the increased production of food for the world's hungry. The second face is turned toward the few remaining farmers who have survived these miracles. It is downcast and silent, like a mourner at a funeral.

Downside to a Revolution

The Green Revolution, a fundamental change in agricultural technology, arose in the 1960s and '70s from the assumption that poverty and hunger in poor countries were the result of low agricultural productivity, that subsistence farming as it had occurred for centuries was the basis of a brutish existence. In response to this assumption, plant breeders hit on an elegant method to increase dramatically the yield of the world's most important crops, especially wheat and rice. Put simply, this plan involved redesigning the plants themselves, increasing the size of the plants' reproductive parts—the seed that we eat—and decreasing the size of the vegetative parts—the stems, roots, and leaves that we throw away. From a technical point of view, this worked. Unfortunately, that's not the end of the story. As in other seemingly simple, technical manipulations of nature, there have been undesirable and unintended consequences.

The primary problem is that Green Revolution agribusiness requires vast amounts of energy to grow and sustain these "miracle crops." Oil must be burned to make the large quantities of nitrogen fertilizer on which these plants depend. Farmers also must invest heavily in toxic herbicides, insecticides, and fungicides [chemicals that kill fungus, or mold]; in irrigation systems; and in spraying, harvesting, and processing machinery for the weakened, seed-heavy plants. Large sums of money must be borrowed to pay for these "inputs" before the growing season starts in the hope that crop sales will allow farmers to repay the debt later in the season. When that hope is frustrated, the farmer often loses his farm and is driven into a migrant pool of cheap labor for corporate-farming operations or is forced to seek work in the landless, teeming cities.

The Green Revolution is an early instance of the co-opting [taking over] of human needs by the technoeconomic system. It is not a black-and-white example: some farmers have been able to keep on farming in spite of the high inputs required; others are mixing traditional methods of farming with selected newer technologies. But the latest manifestation of corporate agriculture, genetic engineering, is black-and-white. Excluding military spending on fabulously expensive, dysfunctional weapons systems, there is no more dramatic case of people having their needs appropriated for the sake of profit at any cost. Like high-input agriculture, genetic engineering is often justified as a humane technology, one that feeds more people with better food. Nothing could be further from the truth. With very few exceptions, the whole point of genetic engineering is to increase the sales of chemicals and bioengineered products to dependent farmers, and to increase the dependence of farmers on their new handlers, the seed companies and the oil, chemical, and pharmaceutical companies that own them.

Fighting Pests—or Helping Them?

Social problems aside, this new agricultural biotechnology is on much shakier scientific ground than the Green Revolution ever was. Genetic engineering is based on the premise that we can take a gene from species A, where it does some desirable thing, and move it into species B, where it will continue to do that same desirable thing. Most genetic engineers know that this is not always true, but the biotech industry as a whole acts as if it were. First, genes are not like tiny machines. The expression of their output can change

> *"The whole point of genetic engineering is to increase the sales of chemicals and bioengineered products to dependent farmers."*

when they are put in a new genetic and cellular environment. Second, genes usually have multiple effects. Undesirable effects that are suppressed in species A may be expressed when the gene is moved to species B. And third, many of the most important, genetically regulated traits that agricultural researchers deal with are controlled by multiple genes, perhaps on different chromosomes, and these are very resistant to manipulation by transgenic technology.

Because of these scientific limitations, agricultural biotechnology has been largely confined to applications that are basically simpleminded despite their technical complexity. Even here we find problems. The production of herbicide-resistant crop seeds is one example. Green Revolution crops tend to be on the wimpy side when it comes to competing with weeds—hence the heavy use of herbicides in recent decades. But many of the weeds are relatives of the crops, so the herbicides that kill the weeds can kill the crops too, given bad luck with weather and the timing of spraying. Enter the seed/chemical companies with a clever, profitable, unscrupulous idea. Why not introduce the gene for resistance to our own brand of herbicide into our own crop seeds, and then sell the patented seeds and patented herbicide as a package?

Never mind that this encourages farmers to apply recklessly large amounts of weedkiller, and that many herbicides have been associated with human sickness, including lymphoma [a kind of cancer]. Nor that the genes for herbicide resistance can move naturally from the crops to the related weeds via pollen transfer, rendering the herbicide ineffective in a few years. What matters, as an agricultural biotechnologist once remarked to me, is earning enough profit to keep the company happy.

> *"The emergence of genetic resistance among ... pests becomes almost a certainty."*

A related agricultural biotechnology is the transfer of bacterial or plant genes that produce a natural insecticide directly into crops such as corn and cotton. An example is Bt (Bacillus thuringiensis), which has been widely used as an external dust or spray to kill harmful beetles and moths. In this traditional use, Bt breaks down into harmless components in a day or two, and the surviving pests do not get a chance to evolve resistance to it. But with Bt now produced continuously inside genetically engineered crops, which are planted over hundreds of thousands of acres, the emergence of genetic resistance among the pests becomes almost a certainty.

Monsanto, one of the world's largest manufacturers of agricultural chemicals, has patented cottonseed containing genes for Bt. Advertised as being effective against bollworms without the use of additional insecticides, 1,800,000 acres in five southern states were planted with this transgenic seed in 1996, at a cost to farmers of not only the seed itself but an additional $32-per-acre "technology fee" paid to Monsanto. Heavy bollworm infestation occurred in spite of the special seed, forcing farmers to spray expensive insecticides anyway. Those farmers who wanted to use seeds from the surviving crop to replace the damaged crop found that Monsanto's licensing agreement, like most others in the industry, permitted them only one planting.

Troubles with Monsanto's genetically engineered seed have not been confined to cotton. In May 1997, Monsanto Canada and its licensee, Limagrain Canada Seeds, recalled 60,000 bags of "Roundup-ready" canola seeds because they mistakenly contained a gene that had not been tested by the government for human consumption. These seeds, engineered to resist Monsanto's most profitable product, the herbicide Roundup, were enough to plant more than 600,000 acres. Two farmers had already planted the seeds when Monsanto discovered its mistake.

Threats to Cattle and People

There is another shaky scientific premise of agricultural biotechnology. This concerns the transfer of animal or plant genes from the parent species into microorganisms, so that the valuable products of these genes can then be produced in large commercial batches. The assumption here is that these transgenic products, when administered back to the parent species in large doses, will simply increase whatever desirable effect they normally have. Again, this is simplistic thinking that totally ignores the great complexity of living organisms and the consequences of tampering with them.

In the United States, one of the most widely deployed instances of this sort of biotechnology is the use of recombinant bovine growth hormone (rBGH), which is produced by placing slightly modified cow genes into fermentation tanks containing bacteria, then injected into lactating cows to make them yield more milk. This is done despite our nationwide milk glut and despite the fact that the use of rBGH will probably accelerate the demise of the small dairy farm, since only large farms are able to take on the extra debt for the more expensive feeds, the high-tech feed-management systems, and the added veterinary care that go along with its use.

The side effects of rBGH on cows are also serious. Recombinant BGH-related problems—as stated on the package insert by its manufacturer, Monsanto—include bloat, diarrhea, diseases of the knees and feet, feeding disorders, fevers, reduced blood hemoglobin levels, cystic ovaries, uterine pathology, reduced pregnancy rates, smaller calves, and mastitis—a breast infection that can result, according to the insert, in "visibly abnormal milk." Treatment of mastitis can lead to the presence of antibiotics in milk, probably accelerating the spread of antibiotic resistance among bacteria that cause human disease. Milk from rBGH-treated cows may also contain insulin [like] growth factor [1], IGF-1, which has been implicated in human breast and gastrointestinal cancers.

Another potential problem is an indirect side effect of the special nutritional requirements of rBGH-treated cows. Because these cows require more protein, their food is supplemented with ground-up animals, a practice that has been associated with bovine spongiform encephalopathy, also known as "mad cow disease." The British epidemic of BSE in the 1990s appears to have been associated with an increased incidence of the disease's human analogue, Creutzfeldt-Jakob disease. There seems little reason to increase the risk of this terrible disease for the sake of a biotechnology that we don't need. If cows stay off of hormones and concentrate on eating grass, all of us will be much better off.

"Owning" Life

Meanwhile the biotechnology juggernaut rolls on, converting humanity's collective agricultural heritage from an enduring, farmer-controlled lifestyle to an energy-dependent, corporate "process." The ultimate co-optation is the patenting of life. The Supreme Court's ruling in the case of *Diamond v. Chakrabarty* in 1980 paved the way for corporations to obtain industrial, or "utility," patents on living organisms, from bacteria to human cells. These patents operate like the patents on mechanical inventions, granting the patent holder a more sweeping and long-lasting control than had been conferred by the older forms of plant patents. The upshot of this is that farmers who save seeds from utility-patented crop plants for replanting on their own farms next year may have committed a federal crime; it also means that farmers breeding utility-patented cattle may have to pay royalties to the corporation holding the patent.

The life patents allowed by the U.S. Patent Office have been remarkably broad. Agracetus, a subsidiary of Monsanto, was issued patents covering all genetically engineered cotton. . . . [These patents were revoked or modified in 1995.] Companies such as DNA Plant Technology, Calgene, and others are taking out patents that cover many recombinant varieties of vegetable species, from garden peas to the entire genus *Brassica,* which includes broccoli, cabbage, and cauliflower. The German chemical and pharmaceutical giant Hoechst has obtained multiple patents for medical uses of a species of *Coleus,* despite the fact that this medicinal plant has been used since antiquity in Hindu and Ayurvedic medicine to treat cardiovascular, respiratory, digestive, and neurological diseases.

> *"In the chaos of technological change, we have lost the distinction between a person and a corporation."*

Somehow, in the chaos of technological change, we have lost the distinction between a person and a corporation, inexplicably valuing profit at any cost over basic human needs. In doing so we have forsaken our farmers, the spiritual descendants of those early Hebrew and Greek farmers and pastoralists [herders] who first gave us our understanding of social justice, democracy, and the existence of a power greater than our own. No amount of lip service to the goal of feeding the world's hungry or to the glory of a new technology, and no amount of transient increases in the world's grain production, can hide this terrible truth.

Genetically Engineered Foods Are Not Safe to Eat

by Martin Teitel and Kimberly A. Wilson

About the authors: *Martin Teitel is the executive director of the Council for Responsible Genetics, a nonprofit organization founded to foster public debate on the implications of genetic technology. Kimberly A. Wilson directs the council's program on commercial biotechnology and the environment. They are the authors of* Genetically Engineered Food: Changing the Nature of Nature, *from which this viewpoint is excerpted.*

Throughout history monarchs have employed food tasters. This rather high-risk line of work was invented not for gastronomic reasons [reasons concerned with the taste of food], but out of a recognition that when we eat food we are placing a great deal of trust in whomever provides that meal. In societies where people grow their own food one has a pretty good idea of the origins of the food, what was sprayed on the crop as it grew, and how it was cooked. When food is produced locally, just keep the peace with the farmer and the chef and you can eat your dinner with no worries.

In nearly all societies nowadays, even monarchies, most people no longer grow their own food. We eat our meals each day with the assumption that what is sold is safe, both because we choose to trust the farmers and food sellers and because we have some degree of faith in government regulators and food inspectors. If this system of trust breaks down, people would understandably be frightened.

Shattered Trust

In the late 1990s, one of the authors was preparing a huge pot of vegetarian chili for a group of friends. Opening a large can of red kidney beans from a local supermarket, he was surprised to find the can full of peeled, white potatoes. Opening another can of beans from the same store, he found the same unexpected contents. Every person at the chili party had the same reaction: "Don't ever shop at that place again." Even though this one error in labeling some cans of beans produced no illness and was presumably an isolated incident, every person seeing the mislabeled cans had the same extreme reaction: their assumptions that food is carefully monitored all the way into their grocery carts were temporarily destroyed.

The genfood [genetically engineered food] industry knows of the tendency of people to have a short fuse when it comes to food safety issues. A Monsanto spokesman told one of the authors that the firestorm of protest over genfoods in Great Britain in 1998 and 1999 was probably strongly propelled by the scare over mad-cow disease in that country in 1996. This statement recognizes that even though mad cow has little to do with genetically engineered food—any more than a mislabeled can of red beans has anything to do with a store's fish or lettuce—when trust in food purity and safety is shattered, people become *very* conservative.

Rise in Food-Caused Illness

And so they should. In the United States, up to 80 million people are estimated to become sick from food-caused illness each year. Nine thousand of them die. Statistics from other countries, when they are available, are comparable. The real incidence of illness is probably much higher: many of us have come home from a barbecue featuring sun-baked potato salad, or tried out a new restaurant that looked a bit seedy, and then attributed that night's sickness to the food we just ate. These relatively minor instances of food-related illness never get reported to the collectors of statistics. We just take some pinkish medicine and wait for the bad time to pass.

Other food-connected illnesses are less benign. By the late 1990s, more than twenty [human] deaths were attributed to mad-cow disease in Great Britain, and the number of people ultimately affected by this slow-moving disease won't be known for some time. A whole class of deadly illnesses that might be related to some food is cancer, possibly arising from the residue of added chemicals on or in the food. Because studies linking slow-onset diseases that have complex causes are still underway, contradictory, or not readily available, some people choose to ignore the possibility of risk until better information is available. Others decide on a more cautious approach and go to the extra trouble and expense of purchasing foods labeled "organic" to avoid consuming those chemicals.

"We eat our meals each day with the assumption that what is sold is safe."

Many food-related illnesses lie somewhere in between a minor bellyache that goes away in a day and tragic death from chemical-triggered cancers. So-called "Chinese restaurant syndrome" comes from ingesting too much monosodium glutamate (MSG), a flavor enhancer used in many prepared and Asian foods.

Because people have significantly varying sensitivities to various foods, some restaurants take the precaution of not adding MSG to any of their food and post signs telling their customers of this choice. As we'll see in this chapter, the great variability of individual reactions to ingested substances—-some people show MSG reactions from relatively small amounts of the additive while others have no discernable symptoms—should lead us to an overall abundance of caution when we are considering food safety policy in general. . . .

Is it safe to eat genfood? Let's look at the information available, and then see what actions a person might take.

Allergies

Each gene contributes a single protein to the genetic "soup" that comprises a living organism. Proteins are crucial substances that play many roles in human physiology. One clear association with proteins involves allergies. When a person exhibits an allergic reaction, what her body is reacting to is a protein, most often a "foreign" or introduced protein.

This leads us to a serious issue that arises in connection to genetically engineered foods. If allergies are associated with introduced proteins, and if genfood is by definition characterized by introduced genes that produce proteins, then we have a situation in which caution about allergies is justified. Allergies have already been proved to pass from one type of food into another via gene transfers. The

> *"The fear of introduced genes triggering dangerous and even fatal allergic reactions is based on sound science."*

fear of introduced genes triggering dangerous and even fatal allergic reactions is based on sound science.

Allergies are common in people, ranging from extreme reactions to exotic fish to a mild sensitivity to airborne pollen in the springtime. The amount of a given protein that might trigger an allergic reaction is highly variable. Some people are allergic to common foods like wheat or eggs, while others are able to eat them with no ill effects. Most of us eat peanuts with pleasure, while a few people can find themselves fighting for their lives when they consume even a miniscule quantity of peanut protein. Because of this, most allergy specialists do not advise patients with food allergies to cut down on the food that they are sensitive to. Instead, they sternly admonish their patients to avoid *any* ingestion of that food in even the tiniest, most insignificant-seeming quantity.

The Protection of Labels

The great solace—and safety—for people with food allergies is the labeling of ingredients. In the United States and in many other countries, food producers are obligated to list the ingredients of any prepared or processed food. People with allergies avoid frantic trips to the emergency room by learning to read package labels carefully. Food manufacturers and distributors avoid costly lia-bility by this same disclosure of contents. Our laws tend to say that if an ingre-dient is revealed, that constitutes fair warning. So if people suffer ill effects from eating something that was properly labeled, they cannot sue the food pro-ducer.

This is not so with genetically engineered food. Even in countries that require genfood labeling, the labels will most often just say that genetically modified substances are in the food container. Because there is no requirement to say which gene has been inserted, people must avoid all genfood if they have aller-gic sensitivities and want to be totally prudent.

It is important to note that, unlike mad-cow disease, there have been no docu-mented cases of deaths due to genfood-caused allergic reactions. However, be-cause an autopsy for a death from allergic shock does not normally test for the presence of genetically engineered food, there is no reliable way of gathering data on genfood allergic reactions.

Genetically modified foods available around the world do not present en-hancements to the buyer and consumer of those foods. The foods do not taste better, provide more nutrition, cost less, or look nicer. Why, then, would a per-son with a food allergy run the risk, however large or small it might be, of a life-threatening reaction when safe alternatives are available?

We just need to make sure that those alternatives *remain* available.

Missing Nutrients

The assumption that many of us would make is that genfood is nutritionally equivalent to nonmodified food. In 1999 the California-based Center for Ethics and Toxics (CETOS) set out to see if this was the case. The people at the center noticed that the research submitted to, and accepted by, the U.S. government to demonstrate the safety of Monsanto's genetically engineered soybeans had been conducted by Monsanto's own scientists. A conflict of interest doesn't necessarily mean that people with the conflict are dishonest, only that, as in this case, their associations automatically put the objectivity of their work into question. The CETOS staff, Britt Bailey and toxicologist Marc Lappé, observed that the soybeans Monsanto tested were not an accurate representation of the soy that appears in stores as food because they were not treated with the herbi-cide Roundup, as they would be in real life. So Bailey and Lappé hired a rep-utable testing firm to conduct tests that would accurately compare Roundup Ready soybeans, treated with Roundup, to conventional soybeans that were identical to the Roundup Ready ones except for the missing engineered gene. The tests were also carefully designed to produce results that reflected real-world conditions. This sort of objectively designed science is what we need to be able to make good decisions about what we buy and eat.

The study was published in a peer-reviewed scientific journal in 1999. The process of peer review is important in science. It means that independent scientists looked at the CETOS study and found it to be based on sound and accept-able methods of scientific investigation. In their study, CETOS found that there was a 12 to 14 percent decline in types of plant-based estrogens called phyto-estrogens. Phytoestrogens are associ-ated with protection against heart dis-ease, osteoporosis (bone loss), and breast cancer. A drop in phytoestro-gens of 12 to 14 percent is a signifi-cant nutritional difference.

> *"Objectively designed science is what we need to be able to make good decisions about what we buy and eat."*

The CETOS study was attacked by the American Soybean Association, whose attack was in turn answered by CETOS. Monsanto also conducted new studies that did not show the same changes in phytoestrogens. The new Monsanto study is difficult to compare with the CETOS study, however, because Monsanto inexplicably used a differ-ent, older method in some of its research. Meanwhile, CETOS stands by its study, which the researchers point out at the very least raises some important questions about nutritional variances in this particular food.

An Unsettled Issue

While scientists sling studies and journal articles at each other, what's a food shopper to do? We can't all be expected to become experts on obscure scientific methods or substances we never knew existed, like phytoestrogens. The govern-ment, which chose to accept the original Monsanto-paid tests as the basis for approval of this food, has been of no use in helping us to make prudent deci-sions.

Further, we have no way of knowing if differences in plant hormones in soy-beans mean much for human health, and more important, if the CETOS study was a fluke or will turn out to be typical if CETOS, a nonprofit organization, can garner sufficient support to conduct more scientific experiments.

What can we conclude about nutrition and genfoods from this example?

Soy products appear in many processed and prepared foods. No one knows just how many, but the words soybean oil, soy flour, isolated soy protein, tex-tured vegetable protein, functional or nonfunctional soy protein concentrates, and textured soy protein concentrates on the label are good tipoffs. As much as 60 percent of all prepared food in a typical U.S. supermarket contains geneti-cally engineered ingredients. Further, many people who do not eat meat rely heavily on soy-based food for important nutritional components of their diet, in-cluding proteins, some fats—and phytoestrogens.

As with allergies and genfood, we are left with more questions than answers. Is genetically modified food more nutritious? There is no evidence for a claim of this type. Is genetically modified food less nutritious? We do not know for certain, yet. Is genetically modified food perhaps more variable in its nutritional value? We have at least one reputable study that suggests yes.

The conclusions we can draw from what is known, and not known, seem to be fairly straightforward. First, there is clearly a great need for many further studies of possible nutritional changes in genetically engineered food, based on the CETOS study. Second, there should be clear, unequivocal labeling of genfood so that people can make their own decisions about their nutrition.

Plants Containing Pesticides

A potential problem arising from herbicide resistant GM crops that is largely being ignored is what is the fate of these chemicals within the plant? Are they stable? If they are degraded, what are the products that are produced? And what health risks do they pose?

Michael Antoniou

It should be no surprise to us that in discussing genetically modified food we need to pay attention to chemicals that are designed to kill plants or animals. Many of the significant genfood crops are engineered to either tolerate higher than usual amounts of herbicides or to contain pesticides inside each cell. Many of the purveyors of genfood are companies that market agricultural chemicals. Engineering plants to require what a company already sells makes business sense to these corporations.

The pesticide most in question is *Bacillus thuringiensis,* Bt. This bacterium was isolated one hundred years ago, although it did not become commercially available in the United States until 1958. While this bacterium is related to a common bacterium that causes food poisoning and is also a close relative of the organism that causes anthrax [a fatal disease], Bt itself is considered relatively safe, especially when compared with synthetic bug-killing chemicals.

> *"Engineering plants to require what a company already sells makes business sense."*

Yet Bt *is* a poison. In its purified form it can be extremely toxic to mammals, including humans, and even in its more usual, nonpurified state there are numerous reports of poisonings and various negative health effects. The chemical may be particularly hazardous to people with compromised [damaged] immune systems. Yet because the EPA has already established that Bt (as a spray *on* crops) is safe, it assumes the toxin is safe to eat *in* crops and does not require testing for human health effects.

When Bt is used by organic growers, it is sprayed on plants. It breaks down rapidly in the environment after killing the target bugs. While there may be health risks to the person applying the pesticide, there is no clear evidence of health problems resulting from people eating food that has been sprayed properly with Bt.

When Bt is engineered into a plant, it may remain present in the plant, and the resulting food, much longer than is the case with conventionally used Bt. There is even evidence that Bt engineered into plants remains after the harvest, so that plant leaves that drop to the ground or plant residue that is plowed under have an effect on the living organisms in the soil.

Possible Health Dangers

What we do not have is a series of clear, independent studies on the long- or short-term health effects of eating food containing the pesticide Bt. According to CETOS, from 1987 through 1998 24 percent of genetically engineered crops released into the environment contained insect-resistant genes. According to this same source, Bt crops are grown in the United States, Brazil, Argentina, China, India, Australia, Canada, South Africa, and Japan. Yet we do not have information in hand to establish the safety of this pesticide for human health. The Environmental Protection Agency (EPA) does not test the plant with Bt in it, it only tests the bacteria in isolation. Essentially, the EPA is not testing the product that humans will be consuming. The Bt toxin produced by the plant and the toxin produced by the bacteria could be different. Until both are properly tested for human health effects, no one will know the effect of eating Bt food crops.

In a case that gained some notoriety in early 1999, a scientist at the Rowett Research Institute in Scotland named Arpad Pusztai tested genetically engineered potatoes on rats. After only ten days the animals suffered substantial health effects, including weakened immune systems and changes in the development of their hearts, livers, kidneys, and brains.

When Dr. Pusztai went public with his findings, he was summarily fired and a commission was convened by his former employers to investigate his work. The commission found Dr. Pusztai's work deficient, yet another panel of twenty independent scientists confirmed both his data and his findings. More recent research by another UK-based [British] scientist showing enlarged stomach walls in rats fed genetically engineered potatoes seems to support Dr. Pusztai, who has stated publicly that he will not eat genfood.

Herbicide-Tolerant Crops

Aside from Bt crops, the other major genetically engineered plant chemical involves herbicide-tolerant plants, primarily the Roundup Ready series of plants from Monsanto. These genetically modified plants include corn and soy as well as oil-producing canola (rape seed) and cotton. As we saw, the plants are engineered to withstand the plant-killing effects of the chemical glyphosate, the main ingredient in Roundup. Monsanto claims that this herbicide tolerance means that farmers can spray the plant-killing poison on their fields more precisely and thus use less, but there are serious concerns about how much herbicide is actually being sprayed.

What about the health effects of herbicide-tolerant crops? Scientists have already linked the herbicides containing glyphosate to cancer. Non-Hodgkins lymphoma, which is one of the fastest-rising cancers in the Western world, increasing 73 percent since 1973, has been connected to exposure to glyphosate and MCPA, another common herbicide.

Since even proximity to such chemicals has been linked to cancer, what are the health risks of eating crops sprayed with glyphosate or genetically engineered with Roundup resistance? While the maker of Roundup insists that when used properly the herbicide is safe, independent studies raise a long list of questions about the long- and short-term health effects of human ingestion of glyphosate.

Bottom line: genetically engineered food hasn't been proved safe. Since wholesome alternatives exist, why not suspend production of genfood until it is shown to be wholesome?

Antibiotic Resistance

As we have seen, some crops are genetically engineered with antibiotic-resistance markers, which enable scientists to test for and track genetically engineered characteristics. Perhaps inadvertently [accidentally], these markers have also been used to find farmers who are allegedly growing genetically modified crops without having paid the license fees. Other substances can be used as markers, but antibiotics are convenient.

The problem is that the use of antibiotic-resistant marker genes in food means that people are eating antibiotic-resistant genes, and perhaps the healthful bacteria in their bodies are taking on resistance to antibiotics along the way. Each year, approximately five billion dollars are spent on the treatment of antibiotic-resistant infections in the United States. The threat of enhanced antibiotic resistance among disease-causing pathogens is more real than ever before as bacteria develop ways to elude even the strongest drugs.

Antibiotics are an effective weapon against disease. However, their efficacy has rapidly fallen with time. For example, in 1952 all cases of staphylococcus infection responded to penicillin treatment. By 1980, only 10 percent of all treatment cases were successful. Overuse of prescribed antibiotics is one of the main culprits behind this increased resistance, as pathogens "learn" to develop chemical resistance mechanisms in response to continuous exposure. As harmful bacteria become more resistant to traditional antibiotics, the need to restrict the unnecessary use of antibiotics is crucial. According to the Genome Therapeutics Corporation, "nearly 9 million people in the US are affected by drug resistant bacterial infections each year and [these infections] are the cause of death for approximately 60,000 of these individuals." Antibiotic resistance among old strains of disease is raising both the cost of treatment and number of deaths each year.

Foods Spur Resistance

While overuse of antibiotics is a serious problem, another growing public health threat is the widespread use of antibiotics in farming. For decades, antibiotics have been administered to livestock to improve the quality and shelf life of meat, eggs, and dairy products. However, this use has not been without harmful consequences. For example, in the 1970s resistant salmonella strains were found in the meat and eggs of chickens treated with antibiotics. Less than a decade later, antibiotic-resistant salmonella that attacked humans had appeared on the scene.

In the face of the misuse of antibiotics in medicine and agriculture, scientists have tried to learn about the origin of antibiotic-resistant genes. There is still a great deal that is not entirely understood. Many genes appear to arise spontaneously, some from mutations in preexisting genes.

Regardless of their origins, antibiotic-resistant genes have displayed a remarkable capacity for spreading between organisms by jumping species barriers. Particularly significant is the fact that bacteria share genes all the time. Robert Havenaar, at the TNO Nutrition and Food Research Institute in the Netherlands, engineered bacteria to contain antibiotic-resistant genes. He found DNA from the bacteria to have a half-life of six minutes in the colon—ample time to get picked up by other bacteria.

> *"Antibiotic-resistant genes have displayed a remarkable capacity for spreading between organisms."*

If antibiotic-resistant genes can indeed jump from plants to pathogenic bacteria, as is suggested but not proved by this research, then we are left with the possibility of human beings finding that their ability to be cured of infections by certain antibiotics could decline or even disappear as people eat food containing antibiotic-resistant genes.

Bovine Growth Hormone

Sold in the United States under the Monsanto brand name Posilac, recombinant bovine growth hormone (rBGH) is injected into cows to increase milk production. Few people would argue that the drug does increase milk production, although in a country that periodically gives away dairy products to deal with the milk surplus, it is difficult to understand why we need even more. Aside from well-documented health problems for the cows, including increases in udder infections, there are a series of health issues for humans.

As early as 1995, at a National Institutes of Health (NIH) conference, the following adverse [harmful] effects of rBGH were identified:

1. Strong role in breast cancer
2. Special risk of colon cancer due to local effects of rBGH on the gastrointestinal [digestive] tract
3. May play a role in osteosarcoma, the most common bone tumor in children, usually occurring during the adolescent growth spurt
4. Implicated in lung cancer
5. Possesses angiogenic properties—important to tumors, some of which se-

crete their own growth factors to promote angiogenesis [growth of new blood vessels]

6. Lastly, the 1995 NIH conference recommended that the acute and chronic effects of IGF-1 [insulin like growth factor 1] be determined in the upper GI tract.

Americans have been drinking milk from cows treated with rBGH since 1993. When the hormone was approved by the U.S. government, the approval was based on studies of rats fed rBGH that showed no toxicological changes. Had there been any such changes, further human studies would have been mandated.

In the 1998 Canadian Gaps Report, we learn that in fact a large proportion of the rBGH-fed rats, between 20 and 30 percent, showed distinct immunological changes, while some male rats showed the formation of cysts of the thyroid and infiltration of the prostate. These are warning signs for possible immune system effects—and possible carcinogenic effects as well.

Understated Risks

The Center for Food Safety and more than two dozen other organizations filed a petition in December of 1998 to reverse Food and Drug Administration (FDA) approval of rBGH/rBST: "We're going to go to the courts and say—you were lied to," said Andrew Kimbrell of the Center for Food Safety. "Essentially it was fraud by the agency and fraud by Monsanto in telling the court that there were no human health effects possible from consuming these products made with rBGH-treated milk. We now know that not to be true."

The Canadian Gaps Report, the banning of rBGH in many countries around the globe, and the findings of a number of studies in the United States and in Europe all point to real, concrete health concerns about bovine growth hormone.

Estimates are that 15 to 30 percent of the milk supply of the United States comes from rBGH-injected cows. Since rudimentary labeling of rBGH milk exists in some communities, including direct labeling as well as the labeling of some milk as "organic," people can avoid feeding their families dairy products containing genetically engineered growth hormones. Where such labeling or alternatives do not exist, there is little choice for people other than . . . becoming active in nationwide efforts to provide people with the option to consume only the food that they feel is safe for their families.

Is Genetically Engineered Food Safe?

We have examined a number of different possible health issues with genetically modified foods. In some instances, such as phytoestrogen decline in genetically engineered soy, or a variety of health questions arising from animal studies of bovine growth hormone, there are ample reasons for people to decide to avoid genetically modified food. In other instances, such as the health effects of ingesting herbicide-tolerant engineered food, there just isn't enough good science yet to be sure.

If the FDA does not require labels, and safety testing is the exception rather than the rule, just what is the U.S. government doing to protect the public? The Hoover Institute's Henry Miller, a fan of genetic engineering, writes, "The FDA does not routinely subject foods from new plant varieties to premarket review or to extensive scientific safety tests." Later he notes that the FDA only follows "the development of foods made with new biotechnology via noncompulsory informal consultation procedures."

The conclusion to all of this is clear. There is no genetically engineered food product on the market now—not one—that is necessary. Each product, which may confer financial benefit to its producers, can be shown to have an alternative that from the consumer's point of view is at least equivalent if not superior. If genfoods do not provide a benefit to consumers, and may be shown to have health hazards now or in the future, why take any risk with your health or your family's health?

Since safety has not been demonstrated and our health is precious, avoid eating all genfood.

Genetically Engineered Crops Will Not Help Poor Farmers

by Peter Rosset

About the author: *Peter Rosset is executive director of Food First/The Institute for Food and Development Policy. He is coauthor of* World Hunger/Twelve Myths *(Grove Press, 2nd ed. 1998).*

When a group of Filipino farmers were asked in late 1999 for their thoughts on genetically engineered rice seeds, a peasant leader responded with what might be called the Parable of the Golden Snail. It seems that rice farmers have long supplemented the protein in their diet with local snails that live in rice paddies. At the time of the Ferdinand Marcos dictatorship [1965–1986], Imelda Marcos [his wife] had the idea of introducing a snail from South America that was said to be more productive and, as such, a means to help end hunger and protein malnutrition. But no one liked the taste, and the project was abandoned. The snails, however, escaped, driving the local snail species to the brink of extinction—thus eliminating a key protein source—and forcing peasants to apply toxic pesticides to keep them from eating the young rice plants. "So when you ask what we think of the new GE rice seeds, we say that's easy," the leader said. "They are another Golden Snail."

Third World governments and farmers are being told over and over again that genetically engineered seeds are being created to end hunger and that they should brush aside other concerns in the name of ending it quickly. Yet the reality is that far more than enough food exists today to provide an adequate diet for every human being on the planet. In fact, overproduction is a leading problem. The truth is that many people are too poor to have access to the abundance around them. Furthermore, the diverse, integrated farming systems found on smaller farms can be far more productive than the uniform monocultures that genetically engineered seeds are designed for. Meanwhile, patents on life allow biotech companies to privatize the genomes of crop varieties bred by farmers, without compensation, while the escape of novel genes through pollen threatens local crop varieties and food security.

Farmers Say No to New Seeds

In late 1999, five hundred angry South Asian farmers took a bus caravan across Europe to dramatize their opposition to genetically engineered crops and the free-trade measures embodied in the World Trade Organization (WTO). They say that WTO regulations allowing transnational corporations to patent germ plasm [inheritable material] from seeds their ancestors have bred over millennia amounts to "bio-piracy." Indian peasant leader Lal Shankar called their struggle "a fight of indigenous agriculture and traditional systems against the North-dominated gene technology and free market." He added, "They are stealing and creating hybrid seeds and then selling them back to us." Indian farmers have singled out biotech giant Monsanto for its heavily hyped public relations claims and full-page ads about ending hunger. In 1998 the Karnataka State Farmers' Association, which claims 10 million Indian peasants as members, announced its "Cremate Monsanto" campaign. Since then they have been publicly burning Monsanto's experimental plots in India. In London, on the caravan, farmer Kumud Chowdury said, "My husband is taking care of our farm, while I am here to kill Monsanto before it kills families like mine."

In Brazil, the powerful Landless Workers' Movement (MST) has made stopping Monsanto soybeans a top priority, vowing to destroy any genetically engineered crops planted in Rio Grande do Sul state, where the sympathetic governor has banned them. Meanwhile, a Brazilian federal court judge has suspended commercial release of Monsanto's soy pending further testing. In Mexico, where the maize plant originated when it was bred in pre-Columbian times from wild relatives, Mexican farmers still maintain 25,000 native varieties. Fears that pollen from imported, genetically engineered corn could contaminate and irreversibly damage this invaluable genetic heritage led the Mexican government to announce a ban on its import from the United States in 1999. Meanwhile Mexico's largest corn-flour company moved to calm consumer fears by eliminating genetically engineered corn from its products.

> *"Far more than enough food exists today to provide an adequate diet for every human being on the planet."*

In Thailand, the government, consumers, environmentalists and even exporters are concerned about genetically engineered crops. Arriving there in mid-October 1999 to attend a conference, I found that such foods were the number-one topic in newspapers, on television and even among people on the streets. On October 13 the Thai government slapped a temporary ban on genetically engineered seeds, just as a controversy exploded over genetically engineered cotton plants being grown illegally by farmers near a Monsanto experimental plot. Thai environmental and consumer organizations immediately voiced their opposition to such crops because of their impact on native biodiversity and food safety. With the new ban in place, the commerce minister suggested that Thailand capitalize on the concerns of European consumers by promoting Thai food exports internationally as free from genetic engineering. Thai cooking oil and seasoning exporters rushed to jump on the bandwagon.

It seems that Filipino peasants are not the only ones concerned about a repeat of the Golden Snail. Growing legions throughout the Third World are asking if they really need these so-called miracle seeds. Certainly the sordid history of technology "altruistically" sent by the North to solve the problems of the South—like large dams that displace thousands, only to silt up rapidly, or pesticides that poison millions, only to become ineffective as pests develop resistance—is enough to make anyone think twice about Golden Snails and magic bullets.

Genetic Engineering Is Cruel to Animals

by Jeremy Rifkin

About the author: *Jeremy Rifkin is president of the Foundation on Economic Trends in Washington, D.C., and a frequent critic of biotechnology. His books include* The Biotech Century *(Tarcher/Putnam, 1998), in which this viewpoint appeared in a somewhat different form.*

The 9 billion animals raised each year by America's meat and dairy industries might be the most forgotten creatures in the nation. Intensively confined, physically deformed, and genetically manipulated, the majority of farm animals feed a nation that is too uncomfortable with their plight to focus on it. In 2000 and beyond, assembly-line treatment promises to worsen, the number of animals promises to grow, and the animal-protection community will face a heightened struggle to reform the way animals are reared and raised for human consumption. . . .

On the eve of the Biotech Century, the new tools of biology are opening up opportunities for refashioning life on earth. Before us lies an uncharted new landscape whose contours are being shaped in thousands of biotechnology laboratories in universities, government agencies, and corporations around the world.

Nowhere is the new genetic science likely to have a more profound impact than on the many animal species that populate our planet. Scientists can now manipulate the genetic information that helps orchestrate the developmental processes of all life forms. Using sophisticated recombinant-DNA and cell-fusion processes, they can insert, delete, and even stitch and fuse together genetic information from unrelated species, creating hybrid forms of life that have never before existed. While the global life-science companies promise a cornucopia of new benefits, a growing number of critics worry that the creation of cloned, chimeric, and transgenic animals could mean untold suffering for our fellow travelers in the animal kingdom.

A Second Genesis

Transgenic animals—those whose genetic material includes genes from another species—are a radical departure from both evolutionary history and classical breeding practices. We have been domesticating, breeding, and hybridizing animals for more than 10 millennia. But in the long history of such practices we have been restrained by species boundaries. Although nature has, on occasion, allowed us to cross closely related species boundaries, the incursions have always been narrowly prescribed. Genetic engineering bypasses species restraints altogether. The implications are considerable and far reaching—as two examples from biotech's earlier days demonstrate.

In the 1980s, Ralph Brinster of the University of Pennsylvania's School of Veterinary Medicine inserted human growth hormone genes into mouse embryos. The mice expressed the human genes and grew twice as fast and nearly twice as big as any other mice. These "supermice," as they were dubbed by the press, then passed the human growth hormone gene on to their offspring. A strain of mice now exists that continues to express human growth genes, generation after generation.

In a comparable feat, scientists in England fused together embryo cells from a goat and a sheep and placed the result into a surrogate animal that gave birth to a sheep-goat chimera, the first such example of the "blending" of two completely unrelated animals in human history.

These results could never have been achieved even with the most sophisticated conventional breeding techniques. In biotech labs, however, the recombinant possibilities are near limitless. This radically new form of biological manipulation changes both our concept of nature and our relationship to it. For the first time in history we become the engineers of life itself. We begin to reprogram the genetic codes of our fellow creatures to suit market objectives. We take on the task of creating a second Genesis, this time a synthetic one geared to the requisites of efficiency and productivity.

Brave New Animals

Today much of the cutting-edge research in animal husbandry is occurring in the new field of "pharming." Researchers are transforming herds and flocks into biofactories to produce pharmaceutical products, medicines, and nutrients in their milk and blood. In 1996 Genzyme Transgenics announced the birth of Grace, a transgenic goat carrying a gene that produces BR-96, an antibody being developed and tested by Bristol-Meyers Squibb to deliver conjugated [attached] anticancer drugs. Companies such as Genzyme hope to produce drugs at half the cost by using transgenic "pharm animals" as chemical factories in the coming years. The company's CEO makes the point that Genzyme's new $10 million facility that makes drugs for Gaucher's disease could be replaced in the near future with a herd of just 12 goats.

The new pharming technology moved a step closer to commercial reality on

February 22, 1997, when Ian Wilmut, a 52-year-old Scottish embryologist, announced the cloning of the first mammal in history—a sheep named Dolly. Wilmut replaced the DNA in a normal sheep egg with the DNA from the mammary gland of an adult sheep. He tricked the egg into growing and inserted it into the womb of another sheep. The birth of Dolly is a milestone event of the emerging Biotech Century. It is now possible to mass-produce identical copies of a mammal, each indistinguishable from the original. Cloning brings society into the bioindustrial age by guaranteeing the kind of quantifiable standards of measurement, predictability, and efficiency that have previously been used to transform inanimate matter and energy into mass-produced material goods.

> *"On the eve of the Biotech Century, the new tools of biology are opening up opportunities for refashioning life on earth."*

While cloning is likely to be a widespread practice in the field of pharming in the coming years, it's also being looked to in the meat industry. The ability to reproduce animals with exacting standards of lean-to-fat ratios and other features would provide a form of strict quality control that has eluded the industry.

Animal clones will also be used to harvest organs for human transplantation. Biotech companies such as Nextran and Alexion are inserting human genes into the germ lines of animal embryos to make their organs more compatible with the human genome and less likely to be rejected. The companies hope to use the new cloning techniques to mass-produce these specially designed animals for spare parts to replace failing human organs. Nextran is conducting tests to see if transgenic pig livers can be used outside the body to help patients with acute liver failure while they wait for a suitable human donor.

A New Form of Cruelty

The public was exposed to the cruel suffering that can result from transgenic animal experiments in a rare film clip aired on American evening news programs in the mid-1990s. Scientists at the U.S. Department of Agriculture's (USDA) research center in Beltsville, Maryland, had microinjected a human growth hormone gene into the genetic code of pig embryos. The goal was to produce pigs that would grow larger and faster and generate increased profits for the livestock industry. The results, however, were quite different from what researchers had anticipated. Several of the animals showed gross abnormalities. One pig in the experiment was arthritic, cross-eyed, and lethargic. "The pitfall that we ran into," says Bob Wall, one of the two researchers on the USDA project, "was that we couldn't control the growth hormone gene in the manner we had hoped we could." The gene USDA researchers inserted either failed to trigger additional growth hormones or "kept the hormone faucet on constantly." Vern Pursel, the lead scientist conducting the transgenic experiments, reported that the muscles in the transgenic pigs degenerated and the animals were so weakened they could barely walk.

Chastened by the adverse publicity over the failed experiment, USDA researchers abandoned the effort and turned their attention to inserting chicken genes into pig embryos to produce pigs with large shoulders. Pursel calls his newest transgenic invention "Arnie Schwarzenegger pigs."

Many of the transgenic animal experiments are designed to increase speed of growth, raise weight, and reduce fat. Critics argue that such experiments inevitably lead to increased stress on the animals, more health problems, and unnecessary suffering. Modern animal-husbandry practices bear witness to the cruel toll inflicted on animals in the interest of increasing profits. For example, the modern broiler chicken has been bred to grow to maturity in less than seven weeks and weigh more than five pounds at the time of slaughter. The animal's legs cannot hold its body weight, and as a result, it suffers from painful leg and foot deformities. Now genetic engineers are seeking a patent in the European Patent Office for a transgenic chicken that contains a bovine (cow) hormone gene. The new transgenic chicken is designed to grow to maturity even faster, with leaner meat and earlier sperm production in males virtually assuring more developmental abnormalities, greater stress, more risk of illness, and increased suffering.

"Muscles in the transgenic pigs degenerated and the animals were so weakened they could barely walk."

Using the new gene-splicing technologies, molecular biologists can even alter the most intimate behavioral characteristics of animals for commercial purposes. At the University of Wisconsin, scientists genetically altered brooding turkey hens to increase their productivity. Brooding hens lay one-quarter to one-third fewer eggs than nonbrooding hens. As brooding hens make up nearly 20 percent of an average flock, researchers were anxious to curtail the brooding instinct because "broodiness" disrupts production and costs producers a lot of money. By blocking the gene that produces the prolactin hormone, biologists were able to engineer a new breed of hen that no longer exhibits the mothering instinct. These hens do, however, produce more eggs.

High Stakes

The question, then, raised by the new biotechnologies and made more immediate and poignant by the sudden creation of clones, chimeras, and other transgenic animals is, How should we regard our fellow species in the coming Biotech Century?

In the Industrial Century just past, many scientists, as well as the general public, came to share Rene Descartes's view of animals as "soulless automata." Now, on the eve of the Biotech Century, a new generation of scientists is beginning to replace the idea of animals as machinery with the idea of animals as information—abstract genetic messages.

The biotech revolution will force each of us to put a mirror to our most deeply held values. Now this mirror is shining on our relationship to our fellow creatures. Very soon it will shine on our relationship to ourselves and our fellow human beings. In the final analysis, how we choose now to define and cast our relationship to other animals will likely come to define, at least in part, how we eventually come to view ourselves and the future of our own species. The stakes, then, are high. We are past due for a spirited public debate about the moral and ethical implications of reducing other animals to manipulable genetic information.

Chapter 3

Is Engineering Human Genes Ethical?

Chapter Preface

Some of the greatest controversies about genetic engineering involve the alteration of human genes. Even some people who see nothing wrong with changing the genes of plants or animals draw the line at altering the human species. They fear that such changes could eventually transform the very definition of humanity—and not necessarily for the better. Supporters of human gene alteration, however, say that it could eradicate presently incurable diseases and might, in time, provide humans with powers presently undreamed of outside of video games and comic books.

Scientists are already changing human genes on a limited basis as a way of treating disease. The first gene therapy took place in September 1990, and several hundred forms of gene therapy are currently undergoing human tests. Supporters say that gene therapy eventually offers as much promise of revolutionizing medicine as the discovery of antibiotics. Critics, however, point out that this form of medical treatment has yet to cure a single patient. They question whether it will ever be safe, effective, or affordable.

Some people fear that if gene therapy does come into widespread use, it will lead to ever-broader definitions of "diseases" to be treated, moving from life-threatening conditions to minor problems or even simple differences from a social ideal. If parents could choose the genes for "designer babies" from a catalog, human diversity might be greatly reduced. Since only the wealthy could probably afford gene enhancement at first, the gap between gene-enhanced and nonenhanced social groups might grow wide enough in time to produce two separate species. Supporters of gene enhancement say that it would eventually benefit everyone. Even before then, it would be no less ethical than parents selecting good schools or other ways of trying to improve their children's chances for health and success.

Changing the genes of individuals causes controversy enough, but arguments become even more intense when the proposed changes would affect those individuals' offspring. This would happen if changes were made in the genes of a person's sex cells (sperm and eggs)—so-called germline genes. Changing germline genes could allow inherited diseases to be wiped out once and for all, say those who find such changes ethical. People who oppose changes in germline genes, on the other hand, warn that a mistake in engineering germline genes could harm not merely a single individual but untold generations.

A final controversy has arisen about human cloning, or the creation of a new individual with exactly the same genes as an existing person. Some people say that cloning would be unethical because it would produce individuals who would in some sense be, or at least be perceived as being, less than human. Others claim that a clone would be nothing more than a delayed identical twin and that cloning could provide a last hope for couples who can have a child in no other way.

As genetic engineering technology improves, more and more alterations of human genes are bound to be suggested. Which ones will be adopted—and what the results will be—remain to be seen.

Gene Therapy Will Improve Human Health

by Jeff Wheelwright

About the author: *Jeff Wheelwright's books include* The Irritable Heart: Gulf War Syndrome and the Limits of Modern Medicine *(Norton, 2001). He is a former science editor of* Life *magazine and a freelance journalist.*

Like a strand of DNA, which coils all over itself within a human cell, the biotechnology industry in San Diego is full of loops and tangles. Biotech, a big part of the "new economy" of the United States, is especially big in this seaside community, where the climate is said to be "technologies-perfect" and the race to put human genes to medical use has brought together large drug companies, shoestring start-ups, university labs and government hospitals.

If San Diego used to stand for bougainvillea, retired admirals and perpetual defense contracts, it now has a street named Genetic Center Drive. The new biotech ventures may have small facilities, and the corporate officers might consist of three shaggy scientists in sandals. But investors are betting that the genes cloned by the companies will deliver new treatments, or even cures, for heart disease, cancer and other ailments.

A Center for Gene Technology

San Diego's biotech nucleus is Torrey Pines Mesa. Located north of downtown, in La Jolla, the mesa is a steep, sandy promontory rising from the Pacific. It is home to both the Salk Institute and the Scripps Research Institute, and at the top of the mesa is the University of California, San Diego (UCSD). Many industries in America have been built upon ideas originating on university campuses. The gene business is unusual in that researchers have been able to patent and commercialize their ideas without relinquishing their university appointments. UCSD itself has become a biotech player, supplementing its government grants with money from pharmaceutical manufacturers.

Stroll through the eucalyptus trees to the UCSD School of Medicine. In the

Center for Molecular Genetics, you'll find the laboratory of Theodore Friedmann, M.D. He heads the university's program in human gene therapy. He also
sits on the federal Recombinant DNA Advisory Committee, which assesses
proposals for clinical trials. Friedmann was in the gene therapy business long
before it was a business—when the idea of manipulating genes to remedy disease was just a concept.

The Start of Gene Therapy

A gene is a message, a biochemical message inherited from your parents. It is
one of 50,000 to 150,000 messages that have helped to shape your body or
make it run. Each gene consists of a section of DNA, that helical polymer that
looks like a twisted ladder. It is actually the rungs of the ladder, composed of
just four molecules that can be arranged in seemingly endless combinations,
that will tell the cell what to do. (In shorthand, the molecules are labeled A, G,
C, and T.) Often a cell is told to produce a protein that will carry out the work
of the body. When a disease results from an absent or insufficient or malformed
protein, the problem usually can be traced to a glitch in the DNA.

All this had been outlined as far back as 1972. That year Ted Friedmann and
another scientist wrote a seminal article, "Gene Therapy for Human Disease?"
alerting the public to the technical and ethical issues on the horizon. There was
a possibility, they wrote, that the "'good' DNA [could] be used to replace the
defective DNA." Fixing the disease at its source in the cell made more sense
than replacing the bad protein over and over, as drug therapies did.

But how would the "good" genes be inserted? The Friedmann paper proposed
using artificial viruses. Because the whole purpose of a virus is to insert its
DNA into a cell and force the cell to work for it, the idea was to make viruses
serve as the carriers, or vectors, of beneficial genes.

Getting Genes into Cells

With the blueprint for gene therapy established, the technology slowly took
form. First, techniques were developed enabling scientists to precisely identify
the DNA lettering of genes. They learned to make copies of segments of the
specific letters, and then to "write" genes and grow them in bacterial cultures.
"Libraries" of genes were developed from which researchers could obtain the
living instructions for both making proteins and making the components of the
viruses that would transport them.

To make a vector, scientists took the genes for the parts of a virus and the
genes for therapeutic proteins and put them, one by one, into "helper cells."
These are special cells from mice, dogs or even humans that are receptive hosts
for the propagation of whole viruses. When all the components are added to
these cells, new gene-carrying viruses are automatically created. By careful cultivation, trillions of the gene-carrying viruses can be produced, harvested and
prepared for eventual clinical use.

Friedmann studied a gene called HRPT. When this gene is not working, the body can't produce a critical enzyme, the lack of which leads to a rare but terrible condition called Lesch-Nyhan. In an experiment in the early '80s, Friedmann and his associates became the first to insert a disease-related gene—the HPRT gene—into cells in which the gene was defective. In lab cultures, the corrected cells produced the critical enzyme. For this and other work, some in the field have called Ted Friedmann, now 65, the father of gene therapy.

But you won't hear him say it, for Friedmann has not yet fixed the disease in human beings. Lesch-Nyhan turned out to be too complex. And widespread gene therapy in general has remained an idea on the horizon, if very much closer today. Since 1990 more than 350 trials have been launched in the United States, enrolling 4,000 subjects, but a proven and permanent treatment of the sick by altering their DNA has not occurred. Over the years that he has lectured on the topic, Friedmann always starts with the gulf between the concept of gene therapy, so glowingly simple, and the thorny implementation, "which will take slow slogging, clinically, until a success." Careful with words, he thinks that "gene transfer" more accurately describes the current state of the field.

Good News and Bad News

The success that Friedmann long anticipated may well have come in spring 2000, two breakthroughs in treatment, actually, although the elation was overshadowed by bad news arriving first. In September 1999, Jesse Gelsinger, a teenager with a metabolic disorder, died during an experiment at the University of Pennsylvania [UPenn]. It appears he had a fatal reaction to a virus bearing therapeutic genes into his system.

Ever since gene therapy moved out of the laboratory and into the clinic, the problem has been to get the desired genes into the right cells in sufficient numbers. In the Gelsinger experiment the target cells were in the boy's liver, where his metabolic disorder was based. The vector used was an adenovirus, of the family of viruses that cause respiratory infections. The DNA of the virus had been crippled so that it could infect the liver cells but not reproduce there. Also, the virus was fitted with the gene for making the missing enzyme. Millions of copies of the vector were injected into a liver artery.

"The problem has been to get the desired genes into the right cells in sufficient numbers."

After Gelsinger's unforeseen reaction, a number of institutions using adenoviruses in humans halted their gene therapy trials, at least temporarily. Then other deaths, in other trials, came to light. The deaths were not necessarily attributable to the vectors, but rather to the serious diseases that had inspired the tests. Still, it was found that several researchers had not reported their "adverse events" to the National Institutes of Health (NIH), as was required. Congress held hearings, and the Food and Drug Administration (FDA), which has ultimate authority, tightened its oversight [regulation] and issued disciplinary actions, though not satisfying everyone. Speaking for many, David Baltimore, president of the California Institute of Technology and an esteemed biologist, asked, "What the hell are we doing putting [adenoviral vectors] in people?" The controversy has worried Friedmann.

The lead investigator at UPenn, James Wilson, was criticized for having his feet in both camps. He stood to gain financially from the outcome of his experiment. Friedmann believes that conflicts of interest should be scrutinized, but that in this instance the fault lay in the team's inexperience. In preliminary trials with monkeys and human volunteers, the adenoviral vector caused liver problems, reactions which may have been misinterpreted. In November 2000, a lawsuit filed by Gelsinger's family was settled out of court.

Success Stories

The two successes overshadowed by Gelsinger's death were the real thing. A team of researchers in Paris reported in April 2000 that they had corrected a severe type of immune deficiency in two infants. First they withdrew bone marrow cells from the children. Then they used a retrovirus to infect the cells with the necessary gene. Put back into the patients, the cells began churning out the protective components of the immune system. "Patients with immune systems fully restored," Friedmann points out. "All tests show the patients normal."

The other report concerned a hemophilia treatment developed by American researchers. In hemophilia the blood fails to clot properly. Patients must receive a clotting factor by means of costly infusions that continue for a lifetime. In type B of the disease, patients lack a protein called Factor IX, controlled by a gene that in them doesn't work. So for the trial a vector modified with the healthy gene was injected directly into the muscles of three subjects. The number of subjects was kept small because the goal was to assess the safety of the technique, not the efficacy. But early results showed that Factor IX was being generated in two patients.

"They [the researchers] reported a correction of a bleeding disorder that has lasted well over a year," notes Friedmann. "A physiological event has occurred, not just the identification of gene product in molecular assays [tests]." In other words, not only was a gene directly transferred into human beings but also it was doing what it was supposed to do.

Making Viruses

In the early 2000s, Friedmann's own research focused on retroviruses. Different from an adenovirus, a retrovirus uses its RNA to get its DNA incorporated into the DNA of its target cells. Friedmann's specialty, and one of the hallmarks of this lab, is creating retroviral vectors, the type most commonly used in gene therapy tests. In incubators and centrifuges whirring just across the hall, retroviruses are being built from scratch. His associates, young Ph.D.s on grants, cut out part of the retroviral machinery and splice in DNA corresponding to a human gene. They also tinker with the virus's outer envelope, sticking proteins onto it that make the virus perform new tricks within the body, such as seeking out specific cells to infect. "Ordinary viruses are promiscuous," he comments. "They go where they want to go."

"The two [gene therapy] successes overshadowed by [Jesse] Gelsinger's death were the real thing."

Friedmann is not overawed by the creation of viruses. To him the retrovirus is a tool. "It's a bag of protein with RNA inside," he says coolly, "a bag with an ability to get inside cells. To make a vector, you can strip down a natural virus or engineer one from the bottom up."

The lab makes "research-grade" vectors. These first novel samples aren't deemed ready to be inserted into human beings. For that you must have a "clinical-grade" vector, one that has passed inspections and met government standards. Friedmann makes reference to "GMP"—good manufacturing practices. The term is taken from the FDA. There is a GMP, clinical-grade, commercial vector lab on the medical school campus, a short walk away. Friedmann and UCSD set it up in 1995 and later secured capital from a foreign drug company.

"Production was needed in large amounts," he says, all of a sudden sounding like a businessman. He opens the door to a nondescript brown building. "UCSD saw the need for good material that you could inject into humans. The idea was that we would spread the clinical use of 'clean' vectors. We concentrated on this aspect to spur the field."

The Gene Business

The company's name is Molecular Medicine LLC. Charles Prussak, the company president, is something of a hybrid. A Ph.D., he started professional life as a research biochemist and saw his DNA taken over, as it were, by the hot virus of the marketplace. His phone ringing constantly, his office walls bare, he relishes the dare and competition of the gene business. "There's a tremendous risk for a tremendous reward," he says. "You're either a bum or a star."

In contrast to the reserved and courtly Friedmann, Prussak, 46, is a rapid-fire guy, humorous but sharp with his elbows. "GMP," he says, gesturing at the forms on his desk, "stands for Generate Mountains of Paperwork. There's a demand for reporting at each step of the [manufacturing] process." Not only is the FDA looking over his shoulder, Roche, the corporate parent, like the rest of the industry, is nervous about vector safety since the death of Jesse Gelsinger. But Prussak, unworried, insists, "One of these products will be out in a few years and we'll have gene therapy."

Molecular Medicine's customers are pharmaceutical and biotech companies for the most part, which are developing their gene therapies in private. Prussak cannot reveal the companies' names. They send in their research-grade vectors, complete with a gene, for scaling up under controlled conditions.

A contract is signed, and the vector goes into production. The lab here is hardly larger than Friedmann's, but bigger equipment fills the space and the production rooms are sealed. To guard against contamination, workers are sampled at each entry for free-riding particles and bacteria. "The purpose of gowning," says Prussak, "is to protect the people from the virus and the virus from the people." The most critical step in the process is at the end, "when we're doing a fill"—putting the vastly multiplied and purified vector into containers for shipment to the customer.

Was one of his customers by any chance Collateral Therapeutics, founded by a UCSD cardiologist named H. Kirk Hammond?

Prussak just smiles.

Help for Sick Hearts

Proceeding 100 yards east on the mesa, you come to the VA San Diego Healthcare System, which is affiliated with the university. Dr. Hammond keeps an office and a lab in the building. Many of the aging veterans being treated here suffer from heart disease. When the arteries become impeded by plaque [fatty deposits], the blood can't get enough oxygen to the heart muscle, especially when the heart is beating fast during exercise or stress. Chest pain called angina results. Angina affects six million Americans, who represent but 10 percent of those with cardiovascular disease. Drugs can relieve angina, and if not, balloon angioplasty, the forced widening of an artery, may clear the blockage. The most invasive but effective operation is to bypass the sick arteries by grafting veins harvested from the leg or redirecting mammary arteries.

Curing coronary artery disease (CAD) is likely to pose a harder challenge for gene therapists than do the rare metabolic disorders in which a single gene is defective and one restored enzyme can turn things around. CAD is "complex, multigenic, multifactorial," [caused by many genes and other factors] observes Friedmann. That genes have a role in its development is incontestable, but lifestyle and diet probably are more important factors.

What if, instead of trying to correct the narrowed or blocked vessels, doctors could help patients to make new ones? The body does this naturally, sprouting small vessels and capillaries, called collaterals, to provide more blood to the vascular regions where blood flow is poor. A heart under stress, even the positive stress of exercise, will make collaterals. Scientists are familiar with the proteins that stimulate the formation of such blood vessels. Called growth factors, the proteins each have a gene regulating their production. In recent years these genes have been located and cloned. Now, what if a diseased heart were fed the means to make extra growth factor? Wouldn't the collaterals and blood flow be increased?

"You're either a bum or a star."

To the bottom line: rival heart researchers, Hammond being one, are trying to show that their patented methods of delivering growth-factor genes will lessen the pain of angina and allow longer periods of exercise.

From Therapy to "Improvements"

Encouraging the growth of collaterals to ease the functional impairment caused by CAD would seem to be a proper medical use of a technique. But it is easy to see that the same skills could be used to enhance the body of a well person. If you look down the road in gene therapy, you will see, not too far off, attempts to change healthy human beings in order to make them better. If the changes are made to sperm or egg cells, the changes may be passed on to subsequent generations. In a word, eugenics is coming. (You might wish to go back to Friedmann's lab and discuss the matter. Eugenics got a bad name because of experiments by the Nazis in the 1930s and '40s, but Friedmann, though a Jew whose family fled the Nazis, is not hostile to eugenics. "You're going to have to face it," he says. "As a public health measure to treat disease, it's OK." As for the possible abuses of eugenics, "it's the social response that's evil," he says, "not the scientific information.")

Teaming Up with Corporations

Kirk Hammond founded Colleratal Therapeutics in 1995 with a few partners. "I sure know a lot more about business than when I started out," he says. "We were five guys and a dog. We had nothing." They had a method to inject a growth-factor gene into the coronary arteries, using a catheter loaded with altered adenoviruses. The next year Hammond published findings that the gene transfer worked in pigs and their heart function improved. In 1996 Collateral Therapeutics landed a major backer in Berlex, the U.S. subsidiary of the German drug giant Schering. In 1998 they went public.

> *"You will see, not too far off, attempts to change healthy human beings in order to make them better."*

He leans across his desk. The 51-year-old Hammond is dark-bearded, intense, with an almost wolfish presence. For a medical school professor to have launched a stock offering was a controversial move among old guard academics on campus, he admits. "You're perceived in a different way, and it was painful, but I stuck with it. I believed that full disclosure of our financial backing was the way to go. The university is not in the business of commercializing the inventions of its inventors. With corporate arrangements, rapid advances are possible—for more than I could do on NIH grants."

Indeed UCSD, holding title to his patents, seems behind Hammond all the way. "There's no loser in the arrangement except to the sheltered academic who fusses about the old days," he says. "I used to be one of those guys until I made a discovery."

Testing for Safety

To minimize the conflict of interest, Hammond did not take an active part in the tests of his technique. It's fine by him that the experiment was farmed out to clinicians around the country, at 12 medical centers in all, although eight patients did have their gene injections here at the VA hospital. Hammond and his associates insisted that a control group be included in the trial. One subject in four got a squirt of saline [salt] solution only, a placebo [an inactive treatment] instead of hundreds of millions of viral particles. Their results were compared with those for the patients who received the genes.

Because the FDA requires that the safety of a new technique be established first, the trial was intended to assess the possible danger of the injections as much as the possible benefit. In early 2001 Hammond is cagey about the outcome, for it hasn't been published in a journal yet. "The treatment was safe, I can say that much," he declares, and the company has announced "positive" preliminary results. According to a doctor involved, some patients

"There are miracles out there."

did show improvement after receiving GENERX, as Collateral named the product. To be exact, some in the trial were able to exercise longer on a treadmill without chest pain halting them, whether it was because of new blood vessels or some other reason. A full-scale test of the effectiveness is planned, the last major hurdle before licensing [being able to sell the product].

Looking Inside the Heart

A few turns down the hall from Hammond's office is the hospital's cardiac catheterization lab, headed by Dr. William Penny. Penny conducts such procedures as angiograms and angioplasty. A catheter, a tube less than two millimeters wide, enters through an artery in the thigh and is threaded up to the region of the heart. A dye is injected that shows up on x-rays. The x-ray pictures show the vasculature [blood vessels] enmeshing the heart and any blockage of it, hence the degree of disease.

Penny plays a videotape of a recent angiogram. The catheter squirts dye like a runaway garden hose around the heart. Parts of the image are blank though, not spidery-looking, which means that the contrast material isn't getting into the smaller veins. The beating muscle is being deprived of blood and oxygen.

"This patient previously had open-heart surgery," Penny explains. He points out the grafted vein, "the only one that's still fully open." The vein disappears into an ominously blank area. Springing off the graft are tiny branches of collaterals.

"Collaterals are nature's way of performing a bypass," remarks the doctor. "Sometimes that's totally sufficient, and other times these new vessels can't cut it. We don't wish to open this man's heart another time. We're trying to get his heart to grow more collaterals.

Chapter 3

An Enthusiastic Patient

A week earlier Penny administered GENERX gene therapy to the man. That is, he had three chances in four of getting it. The subject is in the waiting room right now, as it happens, having come back for his initial checkup. So the trek across the mesa has reached its goal. Not in a medical laboratory or corporate boardroom but at the side of one person in need.

John Smith (not his real name) is a cheerful, ruddy, 65-year-old ex-construction worker, who looks a bit like Santa. "When people say I look good, I say all my scars are on the inside," John jokes. But the picture of health is belied by the delicate way he moves, a watchfulness about his body, lest he strain his tender heart.

Because he was one of the last subjects to enter the trial, it's too soon at any rate for change to have occurred—good changes, more collaterals around his heart, or bad changes, it must be said, if the growth-factor gene were to migrate elsewhere, such as to his retinas [part of his eyes], and produce unwanted capillaries.

All the risks have been explained to him, he says. John has battled angina for 20 years. He was forced to retire at age 51. Two heart bypasses have stripped his legs of veins, and further surgery might not even be possible. "I didn't have anything to lose," he says. "When I heard about [the trial], I was jazzed about it. I said I'll go for it. They can try this stuff out on mice, but the human part is necessary. I'd be helping myself if it worked, and it could be helpful to other people too.

"Yeah, I thought about dying," he adds, "but it's only when you're not feeling well that you think about it. First you're out of breath, then the pain. I'd like to get on a mountain bike again and ride five or six miles.

"Since I've been in the study, I've seen a couple of things on TV about gene therapy. Every time I hear about it, I'm wide-eyed and bushy-tailed. There are miracles out there."

Engineering Germline Genes Is Ethical

by Burke K. Zimmerman

About the author: *Burke K. Zimmerman has taught courses in ethics and human values in science, technology, and medicine at the University of California at Berkeley, George Washington University, and elsewhere. He frequently writes and lectures on bioethical issues as well as heading a biotechnology company in Helsinki, Finland.*

The targeted, fully controlled modification of the human genome in a fertilized embryo is technically feasible. I shall thus begin by assuming, first, that we have a detailed knowledge of the human genome, the functions encoded by each set of genes, and the variations that make us different from one another, and second, that we are able both to correct obvious genetic pathology and to select—without introducing unwanted errors or genetic artifacts—the alleles that confer a variety of known traits to our children.

Parents' Responsibilities

Let us examine the potential uses of such methods in light of the prevailing ethical standards governing the practice of medicine, the autonomy of parents trying to provide the best possible lives for their children, and distributive justice. It is difficult to see how adding germline modification to all of the other things one now can do for one's children, and in fact that one is expected to do to bring them health, quality education, and opportunities for developing their talents, poses anything inconsistent with the ethical norms that prevail, at least in western society.

The physician has an acknowledged ethical responsibility to use whatever methods are available to treat and prevent illness or pathology in his patients. If safe and reliable germline genetic surgery on a newly fertilized embryo is available to correct a known inherited genetic pathology, then, unless other means exist to avert that pathology, the physician has a moral obligation to use this technique to try to ensure the health of the baby. Physicians have no such moral imperative, however, at least according to today's norms, to assist prospective parents in attempts to give their children added genetically determined talents or more desirable physical attributes. But neither, of course, do physicians have an ethical proscription to help people enhance themselves through cosmetic surgery, which is routinely done.

Parents are expected to give their children the best possible opportunities in life. Thus, we promote their health and education and optimize their environments to give them greater opportunities. Why, then, should this responsibility not also ex-

> *"Is there some unwritten social truism that people must forever be bound to play a genetic lottery when they procreate?"*

tend to genetic factors that may determine a child's physical and psychological attributes, including even cognitive ability? While people can slightly improve their children's odds by carefully selecting a mate, only germline intervention will permit full control of their genetic endowments. Is there some unwritten social truism that people must forever be bound to play a genetic lottery when they procreate? Would not the well-accepted principle of autonomy leave such a choice solely to the prospective parents?

Genetic Haves and Have-Nots

Autonomy, however, is often at odds with the principle of distributive justice. Since such germline techniques are not likely to be cheap, at least in the beginning, nor to be covered by national health programs or, in the United States, by private insurance, these techniques would remain, a privilege of the wealthy. It is feared that their use would only widen the gulf between existing social and economic classes. *The Bell Curve,* Herrnstein and Murray's widely attacked work, contends that, for generations, significant selection has been skewing the gene pool between the upper and lower classes. Their arguments may be flawed, but germline methods for the privileged would guarantee such a dichotomy.

Nonetheless, to demand justice in this province when it is not applied elsewhere is inconsistent. In the United States, even minimal healthcare is not available to everyone. The distribution of wealth, privilege, and opportunity, particularly the kind that provides children with a rich environment rather than a culturally and educationally impoverished one, is grossly skewed in most of the world. In the absence of broad and serious commitments to improve distributive justice with respect to the many *existing* elements that contribute to the quality of people's lives, arguments about the inequities of germline intervention ring hollow.

Of course, the techniques of germline intervention, as with computers and other new technologies, will be steadily improved in their reliability, scope, and cost. But if this is the slippery slope, then its effect will be to make this privilege of the wealthy generally available.

I do have some worries, however. The deliberate reassortment or correction of genes that are part of us does not really enter the realm of the unknown. But, one day, someone may be tempted to try to leapfrog human evolution by attempting to design a new gene from supposed first principles. Given our dismal record in predicting the consequences of new technologies, and our perennial smugness in believing we understand far more about nature than we do, I would have grave doubts about the wisdom of such intervention, even with extensive data from animal models. While we may eventually understand how this marvelous creation, including our brain and other components, actually works, we must keep in mind that the human system is the metastable result of a long evolutionary process, and that the pieces all work together in optimized harmony. Adding new or altered components, however good our knowledge of the system, could have unpredictable consequences. This represents a risk that no one should ask an unborn child to assume.

A Difficult Decision

If you could do so safely, would you use an artificial chromosome to extend the lifespan of your child?

Of course I would want my child to have an additional ten years of quality life. Who wouldn't?

I am assuming that, in being offered the opportunity to extend the lifespan of my child-to-be by adding an extra chromosome pair, there were already extensive human data to show that the procedure was safe and actually slowed the aging process. Naturally, I would wish to review personally all of the data and experimental protocols used to establish both the efficacy and the outcome of the procedure. In any case, my decision would be very conservative.

But, while being convinced that the safety and reliability of the procedure would be simply a matter of stringent scientific validation, the question of how an additional chromosome would assort when it comes my offspring's turn to procreate is another matter. My decision would clearly have an important effect, not only on my son's or daughter's life but on his or her children and on all subsequent generations. Therefore, while there may not yet be human data on the next generation, I shall further assume that there are extensive animal data on the fate of such an additional chromosome through-

> *"Adding new or altered [genetic] components, however good our knowledge of the system, could have unpredictable consequences."*

out many generations and on its interaction with the existing set of chromosomes. If my child wishes to have children someday by someone who did not happen to get an extra chromosome, we had better be sure that a dangling unpaired chromosome is not going to cause trouble.

And if, as I was about to allow the procedure to proceed, a last-minute finding indicated a long-term downside of any sort, I would surely change my mind and, no doubt, chastise myself for not having considered such a possibility in the first place. But what if the news comes after his birth?

If I have acted with proper respect for the limitations of the scientific method, then my child should at least understand the basis for my decision. But if I were too conservative, and he found himself aging sooner than his contemporaries who had undergone the procedure, would he resent me one day as being an ultraconservative old fuddy-duddy unwilling to take risks? Unless of course his peers were experiencing an unexpected consequence, for example a much higher than usual cancer rate—would he then be thankful for my wisdom? On the other hand, if I chose to extend his lifespan, he would surely thank me, unless he were one of the excess cancers.

Thus, while I would use my best judgment to do the right thing for my child, nothing is certain. As in all the other decisions we make in bringing up our kids, every choice is something of a crapshoot. There is, therefore, no guarantee whatsoever that the next generation would appreciate my decision, whatever it may be, as every parent of grown children knows all too well.

Cloning Humans Is Ethical

by Gregory E. Pence

About the author: *This viewpoint is excerpted from Gregory E. Pence's book* Who's Afraid of Human Cloning? *Pence is professor of philosophy in the Schools of Medicine and Arts/Humanities at the University of Alabama, Birmingham. He frequently writes about bioethics, and his work has appeared in such publications as* Newsweek, *the* New York Times, *and the* Wall Street Journal.

A clone is not a drone [robot]. Cloned humans would be people. It is a widely-accepted, general principle of modern philosophical ethics that people should be treated equally as moral agents unless there is a morally relevant reason to treat them otherwise. Every person should be treated with respect and as possessing equal moral worth until it is proven that he or she deserves to be treated otherwise.

This principle stems from the acceptance of Kantian [based on the philosophy of Immanuel Kant], utilitarian, and Judeo-Christian theories of ethics that hold that impartiality is central to ethics. As some might say, from the point of view of the universe, no person's suffering should count any less than any other's.

Clones Would Be People

From a practical point of view, acceptance of this principle means that treating people unequally requires passing a test of justification. The onus of proof is on anyone who would treat a cloned human unequally. Anyone who would treat persons of group A unequally from those of group B must specify some morally relevant difference between the two groups. Historically, we have had to learn quite painfully that skin color, religious belief, ethnicity, gender, and sexual orientation are not such differences.

This very strong moral principle entails a sub-principle that society should not discriminate against people according to their origins. "People are people," and it should not matter how they came to exist. Call this the "Principle of Non-Discrimination by Origins."

This is a surprisingly difficult idea for some to understand:

"[Human cloning] would be perhaps the worst thing we have ever thought of in the maltreatment of our species. It would be a kind of new slave class. You would have human beings who were made by other human beings for their purposes." (Nigel Cameron, theologian, bioethicist, and provost, Trinity International University, Deerfield, Ill.)

"It is not at all clear to what extent a clone will truly be a moral agent." (Leon Kass, bioethicist and professor, the University of Chicago)."

If you have a prejudiced reaction to a person, you cannot cite your own prejudice as a moral justification of why that person will be treated badly. This is like a person saying that there should not be racial integration because "those other people" will never accept it.

This principle of non-discrimination by origins means that no one should suffer any prejudice because of how

> *"Society should not discriminate against people according to their origins."*

he was created. Whether a child originated because of unmarried parents, one parent and an unwanted pregnancy, in vitro fertilization, . . . as a twin, triplet, or quadruplet, or quintuplet should not matter. If a child was created as a result of multiple embryo implantation during assisted reproduction, or by cloning, how a child gets into this world does not make him less a person. Instead, we should judge a child by the same criteria that we use to judge any other person.

It took a long time in human history to accept this principle. For millennia, the cultures of Western civilization would not accept children of unmarried women as beings with normal rights: they could not enter synagogues, marry, inherit property, and sometimes, vote. To say that bastards were socially stigmatized is to use a euphemism. Unless you were a king, your illegitimate children were not invited to the important family gatherings at birth, marriage, death, and holiday meals.

Today, we realize that children who were not born to two, married, heterosexual parents had no control over their origins. Once they arrive into the world, such children must be accepted as persons with all the normal rights.

Clones Would Have Human Rights

Legally, humans become persons at birth. The U.S. Constitution has been consistently interpreted in the last fifty years to hold that at birth a human can inherit money, be a tax deduction, and, if killed, be the subject of a homicide charge. Persons originated by cloning would not be slaves or sub-citizens. As law professor Lori Andrews notes, the 13th Amendment would protect them: "Neither slavery nor involuntary servitude, except as a punishment for crime whereof the party shall have been duly convicted, shall exist within the United States, or any place subject to their jurisdiction."

At one time, it was feared that society might treat children born of in vitro fertilization as second-class citizens. Fortunately, children created this way have not been subjected to any special discrimination or ill-regard. Indeed, because they were so desperately wanted, they may be more loved than ordinary children, and the idea that such children would be looked upon by their parents as inferior is laughable. Nevertheless, it took a few minutes to educate the curious: when Lesley Brown took her baby Louise outside after her famous birth, she later said that wary neighbors peered into her baby carriage, expecting to see something "red and wrinkly," like the little monsters in the movie *Gremlins.*

Obviously, many popular ideas about cloned humans are not in the world of real ideas. We have already indirectly disposed of the idea that persons originated by cloning would be slaves, factory drones, automatons, sub-human, or necessarily second-class citizens. Instead, they would just as much be persons as children born from in vitro fertilization. They would be gestated by normal women over nine months. They would be raised by normal parents in normal neighborhoods. The only difference between them and other children is that they would inherit one set of (chosen) genes rather than a randomly mixed set.

Clones Could Not Be Kept as Organ Donors

There is another silly idea that must be dismissed. Because humans originated by cloning are persons, it follows that we cannot kill such persons for their organs. This would be no more ethical or legal than knocking out your brother, transporting him to a hospital, anesthetizing him, and taking out one of his organs for transplantation. Simply because a person is originated in a new way does not mean that, to use Kant's classic terms, he can be used as a "means" to the good of others. Instead, he will be an end-in-himself with the same rights as any other person.

A variation on the theme of spare parts recognizes what bioethicists call the cognitive criterion of personhood. In this imaginative version, the body originated by cloning is never allowed to develop a sense of self or consciousness and is kept alive to be a perfect match for a person whose organs will probably fail one day. Since the cloned humanoid is not a person in the first place, this argument goes, wouldn't it be permissible to keep him alive as a source of future organs for the genetic ancestor or for others?

Here the question has been begged that you could do something to human bodies originated by asexual reproduction that you couldn't do to bodies originated by sexual reproduction. And you certainly cannot stop the development of a normal human baby after birth by giving it a lobotomy and then allowing the brain-injured body to grow to adulthood for organ parts. That is called murder.

To squash this idea from another direction, consider how much outcry there is when a surgeon crosses an ethical line in trying to create more transplantable organs for dying patients on waiting lists. When terminal patients are not resuscitated after a heart attack so that they can be pronounced brain-dead and their organs can be transplanted, *60 Minutes* does an exposé and interviews a local district attorney who threatens to charge the surgeons with homicide if they persist. Suppose physicians somewhere maintained the bodies of adult humans, whose higher brains had never been allowed to develop (at least the brain stem must be kept to regulate organ functions). This would mean excising the cerebrums of sentient human babies at birth. Even if such babies reached adulthood, how could their organs be used for others? There is no way this could be done secretly because organs must be matched by blood type, tissue typing, and size to recipients. Even if we could imagine it being done successfully in secret, the violation of medical ethics would be so horrendous that the exposé would be the equivalent to discovering that infamous Nazi physician Sigmund Rascher was still torturing humans at Ravensbruck in his Sky Wagon.

Don't Call Them Clones

Both Stephen Jay Gould and bioethicist Ruth Macklin have emphasized that the non-people argument is ridiculous because no one today ever considers treating one of a pair of twins as a non-person. Someone may reply by saying that twins or triplets occur naturally, whereas cloned people do not. The premise assumed here is that human worth is maintained if genetic replication occurs randomly, but not if it were deliberately created. If it were deliberate, the resulting beings would be sub-human. So we reach the reductio [ridiculous] conclusion that deliberate creation of a person lessens his worth.

I am going to insist henceforth on one reformative, linguistic stipulation to improve thinking about this topic, at this point in human history and with the legacy about human cloning we have from movies and popular fiction, to refer to people originated by cloning as "clones" or "a clone" is to drastically bias the discussion at the outset in the worst possible way. Popular discussions are full of question-begging phrases, e.g., "to maintain diversity in the human gene pool, it's important that there be no escaping clones" or "if we get some viable clones" or "a clonant . . . could rightly resent having been made a clone" (Leon Kass).

"Clone" connotes sub-human, zombie-like, insect-like behavior. It is associated with phrases such as "an army of clones" and "Slave Clones of Caldor." It is really in the same class as dozens of other nasty terms that slur the racial, ethnic, and sexual origins of a person. So from here on, and despite the cumbersomeness of the phrase, I am going to write about "persons originated by cloning" or even more neutrally, about "nuclear somatic transfer" (NST) to produce a human or about "human asexual reproduction." For the same reason, I will avoid using "to clone" as a verb when it involves humans.

> *"No one today ever considers treating one of a pair of twins as a non-person."*

You Can't Reproduce Yourself

A lot of popular discussion about asexual human reproduction revolves around the question of whether a person could "clone himself." Indeed, most of this discussion assumes that it makes sense to talk this way.

Some of it is funny: "If you have sex with your husband's clone, are you really being unfaithful?" "Would a clone of Bhoutros Bhoutros Ghali's clone be named Bhoutros Bhoutros Bhoutros Ghali?" "If a cloned man and a cloned woman marry, divorce, change partners, and try again, it would be good," said a divorce lawyer. "The same reason the first marriage failed, the second marriage fails. That's four for me instead of two." Of course, to talk this way is to accept uncritically some very questionable assumptions about personal identity.

I [will] discuss some [environmental] reasons why one person originated from the genotype of another would not be exactly identical.

> **"[My clone] would certainly not be an instant, carbon-copy of me."**

Suppose I want to reproduce myself. Suppose I persuade my wife to undergo minor surgery, have an egg removed . . . , have its nucleus cut out and have my genes inserted. She will then spend nine months gestating the embryo, and because I might be killed in some accident, making an implicit commitment to raising the child until adulthood ("no small assumptions!", says my wife in the background as I run this by her). Even if these assumptions came true, would the baby who was born be me? Not likely. He would certainly not be an instant, carbon-copy of me. For one thing (and not to put too fine a point on it), he would be a *baby*, whereas I am 50 years old.

Moreover, he would grow up not in the years after the close of World War II in the suburbs of Washington, D.C., but in the suburbs of Alabama at the beginning of the 21st Century. He would not watch *The Mickey Mouse Club*, buy five-cent Cokes and return the bottle to get back a two-cent deposit. He would not be the oldest of five children, with attendant, later baby-sitting responsibilities. Instead, he would grow up and know about MTV, know how to work a Macintosh computer and get on the Internet, know about the world from CNN, and have to adjust to dogs and cats in his house. More important, he would not have Gil and Louise Pence as his parents but Greg and Pat Pence, and the latter would be different parents than the former (not necessarily better, just different). So if I wanted to clone myself, I would be disappointed.

More Nurture than Nature

We do not know how much nurture contributes to personality over nature. (One thing that cloned humans would help teach us is the answer to this question.) But we believe nurture contributes a lot. Sometimes the contributions of nurture are masked as those of nature.

For example, scientists discovered in 1997 that how much a baby is talked to (and with how much affection) dramatically affects how many neural pathways are formed in its inchoate [undeveloped] brain. Any baby's brain is like a wild garden, where talking forms pathways that cannot later be formed. Once the garden grows too wild, it cannot later be tamed. By using NMR (nuclear magnetic resonance) scans of the brains of such children at age two, scientists were able to document huge gains or losses from the results of the verbal activities of parents. All of these pathways seem to get laid down by the age of two, and scientists found huge differences in how much they talked to their babies between (to use their categories) parents who were on welfare, who had blue collar jobs, and who were professionals.

So the interventions of parents may be much more important than genetic reductionism [the belief that most personality is determined by genes] implies. The ultimate constitution of the phenotype, the observable physical traits of a specific adult, is in an ongoing, evolving interaction with a changeable, variable environment. And here the environment matters a great deal. A person originated from Hitler's genes could end up a rabbi. A person cloned from Michael Jordan's genes could be a television weatherman who couldn't care less about basketball. A genotype originated from Placido Domingo might only be an average singer. Remember, milk from nine cows produced from nine twinned embryos is only predicted to be 70% similar.

The Mother's Role

Finally, for a male to talk about cloning himself is really far-fetched. The only way he is going to do so is to find a willing woman to gestate the embryo with his replicated genes, and whoever that woman is, will be the gestational mother and contribute her mitochondrial DNA to the resulting child. Unless surrogacy is involved (not a likely prospect), that woman will also be the real, rearing mother, and hence, will contribute a lot to the development of that boy.

So if a man wants to replicate himself, he better think it through. It is likely that his relationship, or the lack of it, in the daily activities of raising a child will have as much to do with whether the child is like him than any connection he has to the child through his genes.

Will rich men hire surrogate mothers to gestate and raise their cloned embryos? Not too often. Only three American states currently legalize commercial surrogacy, i.e., they consider surrogacy contracts enforceable in court. Many states ban such contracts altogether. In any case, surrogacy is expensive and a lot of things can go wrong. A woman employed as a surrogate always has a legal right to abort. As the Baby M case taught us, the surrogate

> *"The requirement of informed consent to enter existence is preposterous."*

mother may bond with the baby and change her mind about giving up the baby.

However, I suppose that if a rich person were fanatical enough about originating a genetic copy of himself, he could attempt it. I don't see anything that would prevent him from flying to some offshore island with various fertility specialists. He could hire local women to gestate his genes in a re-implanted enucleated [with the nucleus removed] egg. He could give a huge bonus for actual birth, say five thousand U.S. dollars, encouraging any pregnant woman to take all possible precautions to avoid miscarriage.

Even then, under the law of whatever country he was in, the mother of the child would undoubtedly be the legal parent. Whether he could adopt the child would be up to the mother and the laws of that country, and it is certainly not a sure thing that all this would work out. The mother might bond to the child, change her mind, blackmail him for more money, or want to get married. Each woman who gestated his child might sue him in U.S. courts for child-support or other monies. Even if he adopted the child, he would need to wait many years, perhaps twenty, to see how closely the child resembled him. Since many rich men only get such fanatical ideas late in life as they see their death coming, it is unlikely that he would ever know if his experiment was successful.

> *"Misleading ideas abound about creating a human by cloning."*

Lack of Informed Consent Doesn't Matter

Another misconception can be dismissed very quickly. . . . George Annas invoked an objection of conservative Christian fundamentalist Paul Ramsey, who argued in the early 1970s against all forms of in vitro fertilization because it constituted "non-therapeutic experimentation on the unborn without their consent." After the announcement of Dolly's cloning in 1997, religion columnist Mike McManus echoed Ramsey's criticism, implying that because a child would never be able to give informed consent to being brought into existence through cloning, to do so would be unethical. Similarly, George Annas argued that because experimentation on humans without informed consent is fundamentally unethical, embryonic experimentation is therefore unethical because embryos can never consent to such experimentation: "The birth of a human from cloning might be technologically possible, but we could only discover this by unethically subjecting the planned child to the risk of serious genetic or physical injury, and subjecting a planned child to this type of risk could literally never be justified."

The argument is so silly that it would not be worth mentioning but for the fact that otherwise sane people continue to use it. First, the requirement of informed consent to enter existence is preposterous. ("Charles, dear, please consent to be born. Look, it really isn't so bad out here. Let me tell you why. . . .") One has to be a person to give consent and one can't give consent without a being a person, so one can't consent to become a person.

Second, if IVF or cloning is the only way that a child is going to exist, and if there is every expectation that the child will be normal, then we can use something that researchers call presumed consent. This kind of consent is where we predict what reasonable people would consent to in experiments where actual consent is impossible (e.g., because some deception is needed that prevents consent). So here the question of presumed consent is: "Would you be willing to take a tiny risk at birth of defects in order to exist in the first place?" And the answer to that is obvious. . . .

Reproductive Freedom Doesn't Lead to Coercive State Eugenics

There is an invidious association that different forms of originating humans lead to eugenics which in turn lead to state-controlled breeding and the destruction of reproductive liberty in ordinary couples.

Some of this is due to not understanding the point of the famous classic, *Brave New World,* by Aldous Huxley. Written in 1932, this novel has been often cited by social critics as having predicted the kind of results they now fear—a future in which governments use technology to control reproduction.

This view of *Brave New World* is mistaken. The controls Huxley had imagined were based on psychological conditioning and need to be seen in the context of behaviorism, a school of psychology which was then as feared and misunderstood as cloning seems to be now. It is mistaken to extend Huxley's fictional ideas about psychological manipulation to fears about real biological innovations.

The great irony of all these citations is that in *Brave New World,* Huxley described the devastating consequences of loss of reproductive choice by individuals. Arguments that human cloning should be banned because of the message of *Brave New World* amount to saying that what the novel opposed should occur, that couples should be denied reproductive choices.

In conclusion, some misleading ideas abound about creating a human by cloning. Many of them come from science fiction. Our history of false alarms about creating babies through in vitro fertilization should make us cautious about bans over any kind of human reproduction. Some of these misconceptions are so unreal as to be funny.

Gene Therapy Is Dangerous to Health

by Jeremy Rifkin

About the author: *Jeremy Rifkin is president of the Foundation on Economic Trends in Washington, D.C., and a frequent critic of genetic engineering. His books include* The Biotech Century *(Tarcher/Putnam, 1998) and* The Age of Access *(Tarcher/Putnam, 2001).*

It's been more than a decade since I appeared before the RAC [the US National Institutes of Health's Recombinant DNA Advisory Committee] in 1988. At that time, RAC was considering approving the first gene therapy experiment. I warned this committee that proposed somatic gene therapy experimental protocols, using viral vectors, posed serious potential health risks to patients. I also argued that the pre-clinical research was far too scanty and inconclusive to warrant human gene therapy trials. Finally, I warned the RAC that the commercial ties of researchers involved in these trials raised the specter [disturbing possibility] that financial considerations might create an inherent conflict of interest, undermine their objectivity, and bias their reporting of results, including the reporting of adverse effects. The RAC and the scientific community said our concerns were alarmist and unfounded.

Now, in 2001, we face the grim reality of several unaccounted for deaths and the reporting of hundreds of adverse events using adenoviral vectors. In addition, researchers have failed to disclose vital information to government agencies, and biotech companies have hidden behind the veil of trade secrets, patent protection and proprietary information. In the same ten year period, more than 300 separate experiments involving thousands of patients have yet to result in a single cure. The only demonstrable effect of all of this experimentation is that biotech companies have secured substantial investment capital and increased the value of their stock, making small fortunes for their principals, including the researchers involved in the experiments.

It's now time to re-evaluate a field of research that's run amok [gotten out of control] before any more people die needlessly. Today, I am proposing an amendment to the NIH guidelines that would place an immediate moratorium on the continuation of any somatic gene therapy trials using adeno, retro or other viral vectors, except in cases of life and death, last resort. The Foundation on Economic Trends is seeking a RAC-initiated action that could temporarily halt the FDA review and approval process for all protocols using viral vectors, or at least an action that could halt NIH approval of NIH-funded protocols using viral vectors.

Protecting Patients

The Foundation on Economic Trends (FET) is quite aware of the fact that 89% of the gene transfer clinical trials currently registered with the NIH are Phase I studies designed to assess safety and toxicity of the gene transfer intervention and that only 11% of the trials focus on efficacy. It is for this very reason that the foundation is proposing that somatic gene therapy employing viral vectors be limited for now to only terminal patients. This would allow the research to continue and additional data on safety and toxicity to be gathered, without placing other non-terminal patients at risk of adverse effects, including death.

If the additional data collected in these trials leads to a greater warranty regarding safety and toxicity, the Foundation suggests a phase down moratorium, allowing viral vector somatic gene therapy transfers to be used on patients regarded as ill, but less than terminal. Data on safety and toxicity gathered in these trials, in turn, would be used to justify a further moratorium phase down to the next category of illness.

Because the vast majority of these somatic gene therapy trials are designed merely to test for safety and toxicity and not for efficacy—according to the RAC—and because of the revelations of hundreds of additional adverse reaction reports and several unexplained deaths, it would be cruel and irresponsible to subject non-terminal patients to serious health risks knowing that no appreciable benefit is likely to occur to the patient.

The RAC argues that if trials were to be limited only to treatment of life threatening illnesses, that it would be "difficult to make sense of the data gathered." FET believes that the difficulty in assessing the data needs to be weighed against the prospect of subjecting healthier patients (with non-life threatening illnesses) to potential risks, including death. The fact is, after 10 years of human trials, adverse effects continue to mount, more patients are dying, more questions about safety and toxicity are being raised, and many researchers acknowledge that they are no closer to finding the answers. The only thing we know for sure is that corporate portfolios are being hyped and money is being made in anticipation of

> *"It's now time to re-evaluate a field of research that's run amok before any more people die needlessly."*

cures that are as elusive today as they were when these trials began more than a decade ago.

The RAC makes the point that other non-viral vectors currently used in somatic gene therapy trails might pose potential risks to patients as well. We concur and are currently reviewing the newly collected data on these trials to assess whether a similar phase down moratorium might be applied in these trials as well.

It's time for considered reason to prevail over false hopes and for patient safety to be placed above corporate interest. Even one more unnecessary death in these gene therapy trials is one death too many. In the final analysis, biotech research will be better served if we implement a phase down moratorium on the use of viral vectors in somatic gene therapy trials.

Engineering Germline Genes Is Unnecessary and Unethical

by Ruth Hubbard and Elijah Wald

About the authors: *Ruth Hubbard is professor emerita of biology at Harvard University. She writes frequently about the politics of health care, and her works include books such as* The Politics of Women's Biology. *Elijah Wald is a freelance writer and musician. This viewpoint is excerpted from their book* Exploding the Gene Myth *(Beacon Press, 1997).*

Attempts to modify the DNA in our reproductive cells (sperm and eggs) or in the cells of early embryos raise . . . troubling issues. If the DNA of these cells is altered, this will not just affect a specific individual. . . . The altered DNA will be passed to future generations in the *germ line,* which is why this type of DNA manipulation is conventionally called *germ-line gene therapy.* Let me make clear how it differs from [other] sorts of gene manipulations. . . .

Altering the Germ Line

If the DNA in a differentiated tissue, such as liver or skin, is altered by inserting or modifying all or part of a functional DNA sequence, this will only affect the person whose tissue is being modified. So, if it does no good or, worse yet, does harm, only that person will suffer the consequences. However, if one modifies the DNA in a sperm or egg or in the cells of an early embryo, the altered DNA will be copied each time these cells divide and will become part of all the cells in that future person, including her or his eggs or sperm.

This sort of manipulation is not intended to treat the health problems of actual people, but rather to alter the genetic makeup of hypothetical future people. Current attempts at germ-line gene manipulation involve the use of very early embryos, produced in a dish by in vitro fertilization. When the fertilized egg has divided into six or eight cells, the scientists remove one or two cells and test them to see whether they have the mutation the scientists are trying to remedy. Because at this stage all of the cells are still equivalent, removing a couple of them does not damage the embryo. If the embryo has the suspected mutation, the scientists could try to correct it through gene manipulation, before inserting it into the womb of the woman who will carry the pregnancy.

A Better Way to Prevent Disease

This is interesting science, but it is hard to see whom it would help. In the in vitro procedure, at least a half-dozen eggs are fertilized, of which only a few are usually implanted. The rest may be frozen and stored for later use, or discarded. Since there is a choice of which embryo to implant, one could select those which do not have the mutation; there is no reason to choose those which do have the mutation and then attempt to correct it. The one situation in which such embryo selection would not work is if all of the embryos had the mutation. This will only happen in the rare situations in which both parents have two copies of the same recessive allele, as for example if they both have sickle-cell anemia or the same form of cystic fibrosis. If such couples do not want a child with the same condition, there is a range of available options, from adoption to using a sperm donor. It hardly seems reasonable to develop germ-line gene manipulation for this purpose.

In terms of curing or treating genetic conditions, germ-line manipulations are completely irrelevant. There are no sick people who will benefit, and there are other ways to avoid passing on specific genetic traits. What is more, this technology could have some frightening consequences. As we have seen, if scientists alter the DNA of an early embryo those alterations will not only be incorporated into the cells of the person into whom that embryo may develop, but into the cells of her or his children, becoming a permanent part of the hereditary line.

Permanent Mistakes

This permanence raises troubling questions. It is not unusual for medical therapies to have unanticipated, undesirable effects. These are commonly called "side effects," though they may have more serious consequences than the intended effects. If a treatment produces an actual disease condition, that disease is called *iatrogenic*, which is Greek for "medically generated."

Iatrogenic conditions can be serious, even deadly, but often if a medication evokes untoward symptoms it

> *"[Germline gene alteration] is interesting science, but it is hard to see whom it would help."*

can be stopped and, with luck, the symptoms will disappear. If ill effects show up in the next generation, as with DES or thalidomide, this is still not a permanent genetic change. However, if a genetic manipulation of the germ line turns out to be iatrogenic, medical practitioners will have become sorcerer's apprentices. The condition they have introduced will be beyond their control and it will be heritable.

By saying this, I do not want to overstate the potential significance I am prepared to assign to genes. I have said repeatedly that genes may turn out to play a considerably less important metabolic role than molecular biologists and geneticists tend to think they do. But the range and variety of the effects that may result from inserting or modifying chunks of DNA are unpredictable in any specific instance.

> *"Tampering with DNA will have unexpected effects, and there is every reason to believe that some of them will be undesirable."*

We must remember that a single gene may function differently in different tissues. The fact that scientists have linked a DNA sequence with a specific trait does not mean that the sequence has no other functions. It may participate in other metabolic reactions that scientists know nothing about. Tampering with DNA will have unexpected effects, and there is every reason to believe that some of them will be undesirable.

From Therapy to Enhancement

To introduce changes into an individual's hereditary line goes way beyond what we ordinarily think of as a justifiable medical intervention. Yet, if the DNA of early embryos proves to be easier to manipulate than DNA in the differentiated tissues of children or adults, some scientists will advocate manipulating the DNA in germ cells. If attempts at somatic gene manipulation have been less successful than promised, germ-line manipulations can be touted as a more effective way to get the same results, not for people with the conditions, but for their future children. However, if somatic gene manipulations are successful in some cases, people may be persuaded that germ-line manipulations are a logical next step.

As Edward Berger, a biologist, and Bernard Gert, a philosopher, point out:

> Past experience has shown that exciting new technology, including medical technology, generates pressures for its use. Thus, it is quite likely that if germ-line gene therapy were allowed, it would be used inappropriately. . . . In the real world researchers will overestimate their knowledge of the risks involved and hence will be tempted to perform germ-line gene therapy when it is not justified.

Lest this sound unduly alarmist, here is a quotation from Daniel Koshland, a molecular biologist and editor-in-chief of *Science* magazine. Writing on the ethical questions posed by germ-line gene manipulations, Koshland muses about the possibility "that in the future genetic therapy will help with certain types of IQ deficiencies." He asks, "If a child destined to have a permanently low IQ could be cured by replacing a gene, would anyone really argue against that?" (Note the use of the word "cured" for averting the "destiny" of a "child" who would, at the time of the cure, be a half dozen cells in a petri dish.) While voicing some misgivings, Koshland continues:

> It is a short step from that decision to improving a normal IQ. Is there an argument against making superior individuals? Not superior morally, and not superior philosophically, just superior in certain skills: better at computers, better as musicians, better physically. As society gets more complex, perhaps it must select for individuals more capable of coping with its complex problems. . . .

Dangerous Decisions

Clearly, the eugenic implications of this technology are enormous. It brings us into a Brave New World in which scientists, or other self-appointed arbiters of human excellence, would be able to decide which are "bad" genes and when to replace them with "good" ones. Furthermore, the question of whether to identify the functions of particular genes or to tamper with them will not be decided only—or perhaps even primarily—on scientific or ethical grounds, but also for political and economic reasons. We need to pay attention to the experiments that will be proposed for germ-line genetic manipulations, and to oppose the rationales that will be put forward to advance their implementation, wherever and whenever they are discussed.

Cloning Humans Is Not Ethical

by Leon R. Kass

About the author: *Leon R. Kass is Addie Clark Harding Professor in the Committee on Social Thought and the College of the University of Chicago. He has written several articles opposing human cloning and has coauthored (with James Q. Wilson) a book on the subject,* The Ethics of Human Cloning *(AEI Press, 1998).*

"To clone or not to clone a human being" is no longer a fanciful question. Success in cloning sheep, and also cows, mice, pigs, and goats, makes it perfectly clear that a fateful decision is now at hand: whether we should welcome or even tolerate the cloning of human beings. If recent newspaper reports are to be believed, reputable scientists and physicians have announced their intention to produce the first human clone in 2001 or 2002. Their efforts may already be under way.

The media, gawking and titillating as is their wont, have been softening us up for this possibility by turning the bizarre into the familiar. Since the birth of Dolly the cloned sheep in 1997, the tone of discussing the prospect of human cloning has gone from "Yuck" to "Oh?" to "Gee whiz" to "Why not?" The sentimentalizers, aided by leading bioethicists, have downplayed talk about eugenically cloning the beautiful and the brawny or the best and the brightest. They have taken instead to defending clonal reproduction for humanitarian or compassionate reasons: to treat infertility in people who are said to "have no other choice," to avoid the risk of severe genetic disease, to "replace" a child who has died. For the sake of these rare benefits, they would have us countenance [accept] the entire practice of human cloning, the consequences be damned.

Stopping a Dangerous Trend

But we dare not be complacent about what is at issue, for the stakes are very high. Human cloning, though partly continuous with previous reproductive technologies, is also something radically new in itself and in its easily foresee-able consequences—especially when coupled with powers for genetic "en-hancement" and germline genetic modification that may soon become avail-able, owing to the recently completed Human Genome Project. I exaggerate somewhat, but in the direction of the truth: we are compelled to decide nothing less than whether human procreation is going to remain human, whether chil-dren are going to be made to order rather than begotten, and whether we wish to say yes in principle to the road that leads to the dehumanized hell of *Brave New World*.

In 1997 I addressed this subject in an essay, trying to articulate the moral grounds of our repugnance at the prospect of human cloning. Subsequent events have only strengthened my conviction that cloning is a bad idea whose time should not come; but my emphasis this time is more practical. To be sure, I would still like to persuade undecided readers that cloning is a serious evil, but I am more interested in encouraging those who oppose human cloning but who think that we are impotent to prevent it, and in mobilizing them to support new and solid legislative efforts to stop it. In addition, I want readers who may worry less about cloning and more about the impending prospects of germline genetic manipulation or other eugenic practices to realize the unique practical opportunity that now presents itself to us.

For we have here a golden opportunity to exercise some control over where biology is taking us. The technology of cloning is discrete and well defined, and it requires considerable technical know-how and dexterity; we can there-fore know by name many of the likely practitioners. The public demand for cloning is extremely low, and most people are decidedly against it. Nothing sci-entifically or medically important would be lost by banning clonal reproduc-tion; alternative and non-objectionable means are available to obtain some of the most important medical benefits claimed for (non-reproductive) human cloning. The commercial interests in human cloning are, for now, quite limited; and the nations of the world are actively seeking to prevent it. Now may be as good a chance as we will ever have to get our hands on the wheel of the run-away train now headed for a post-human world and to steer it toward a more dignified human future.

What Is Cloning?

What is cloning? Cloning, or asexual reproduction, is the production of indi-viduals who are genetically identical to an already existing individual. The pro-cedure's name is fancy—"somatic cell nuclear transfer"—but its concept is simple. Take a mature but unfertilized egg; remove or deactivate its nucleus; in-troduce a nucleus obtained from a specialized (somatic) cell of an adult organ-ism. Once the egg begins to divide, transfer the little embryo to a woman's uterus to initiate a pregnancy. Since almost all the hereditary material of a cell is contained within its nucleus, the re-nucleated egg and the individual into which it develops are genetically identical to the organism that was the source of the transferred nucleus.

An unlimited number of genetically identical individuals—the group, as well as each of its members, is called "a clone"—could be produced by nuclear transfer. In principle, any person, male or female, newborn or adult, could be cloned, and in any quantity; and because stored cells can outlive their sources, one may even clone the dead. Since cloning requires no personal involvement on the part of the person whose genetic material is used, it could easily be used to reproduce living or deceased persons without their consent—a threat to reproductive freedom that has received relatively little attention.

Some possible misconceptions need to be avoided. Cloning is not Xeroxing: the clone of Bill Clinton, though his genetic double, would enter the world hairless, toothless, and peeing in his diapers, like any other human infant. But neither is cloning just like natural twinning: the cloned twin will be identical to an older, existing adult; and it will arise not by chance but by deliberate design; and its entire genetic makeup will be pre-selected by its parents and/or scientists. Moreover, the success rate of cloning, at least at first, will probably not be very high: the Scots transferred two hundred seventy-seven adult nuclei into sheep eggs, implanted twenty-nine clonal embryos, and achieved the birth of only one live lamb clone.

Reactions to Cloning

For this reason, among others, it is unlikely that, at least for now, the practice would be very popular; and there is little immediate worry of mass-scale production of multicopies. Still, for the tens of thousands of people who sustain more than three hundred assisted-reproduction clinics in the United States and already avail themselves of in vitro fertilization and other techniques, cloning would be an option with virtually no added fuss. Panos Zavos, the Kentucky reproduction specialist who has announced his plans to clone a child, claims that he has already received thousands of e-mailed requests from people eager to clone, despite the known risks of failure and damaged offspring. Should commercial interests develop in "nucleus-banking," as they have in sperm-banking and egg-harvesting; should

> *"We have here a golden opportunity to exercise some control over where biology is taking us."*

famous athletes or other celebrities decide to market their DNA the way they now market their autographs and nearly everything else; should techniques of embryo and germline genetic testing and manipulation arrive as anticipated, increasing the use of laboratory assistance in order to obtain "better" babies—should all this come to pass, cloning, if it is permitted, could become more than a marginal practice simply on the basis of free reproductive choice.

What are we to think about this prospect? Nothing good. Indeed, most people

are repelled by nearly all aspects of human cloning: the possibility of mass production of human beings, with large clones of look-alikes, compromised in
their individuality; the idea of father-son or mother-daughter "twins"; the
bizarre prospect of a woman bearing and rearing a genetic copy of herself, her
spouse, or even her deceased father or mother; the grotesqueness of conceiving
a child as an exact "replacement" for another who has died; the utilitarian creation of embryonic duplicates of oneself, to be frozen away or created when
needed to provide homologous tissues or organs for transplantation; the narcis-
sism of those who would clone themselves, and the arrogance of others who
think they know who deserves to be cloned; the Frankensteinian hubris to create a human life and increasingly to control its destiny; men playing at being
God. Almost no one finds any of the suggested reasons for human cloning compelling, and almost everyone anticipates its possible misuses and abuses. And
the popular belief that human cloning cannot be prevented makes the prospect
all the more revolting.

Revulsion is not an argument; and some of yesterday's repugnances are today
calmly accepted—not always for the better. In some crucial cases, however, repugnance is the emotional expression
of deep wisdom, beyond reason's
power completely to articulate it. Can
anyone really give an argument fully
adequate to the horror that is father-
daughter incest (even with consent),

> *"Most people are repelled by nearly all aspects of human cloning."*

or bestiality, or the mutilation of a corpse, or the eating of human flesh, or the
rape or murder of another human being? Would anybody's failure to give full
rational justification for his revulsion at those practices make that revulsion ethically suspect?

I suggest that our repugnance at human cloning belongs in this category. We
are repelled by the prospect of cloning human beings not because of the
strangeness or the novelty of the undertaking, but because we intuit and we
feel, immediately and without argument, the violation of things that we right-
fully hold dear. We sense that cloning represents a profound defilement of our
given nature as procreative beings, and of the social relations built on this natural ground. We also sense that cloning is a radical form of child abuse. In this
age in which everything is held to be permissible so long as it is freely done,
and in which our bodies are regarded as mere instruments of our autonomous
rational will, repugnance may be the only voice left that speaks up to defend the
central core of our humanity. Shallow are the souls that have forgotten how to
shudder.

Yet repugnance need not stand naked before the bar of reason. The wisdom of
our horror at human cloning can be at least partially articulated, even if this is finally one of those instances about which the heart has its reasons that reason
cannot entirely know. I offer four objections to human cloning: that it constitutes

unethical experimentation; that it threatens identity and individuality; that it turns procreation into manufacture (especially when understood as the harbinger [warning sign] of manipulations to come); and that it means despotism over children and perversion of parenthood. Please note: I speak only about so-called reproductive cloning, not about the creation of cloned embryos for research. The objections that may be raised against creating (or using) embryos for research are entirely independent of whether the research embryos are produced by cloning. What is radically distinct and radically new is reproductive cloning.

Unethical Experiments

Any attempt to clone a human being would constitute an unethical experiment upon the resulting child-to-be. In all the animal experiments, fewer than two to three percent of all cloning attempts succeeded. Not only are there fetal deaths and stillborn infants, but many of the so-called "successes" are in fact failures. As has only recently become clear, there is a very high incidence of major disabilities and deformities in cloned animals that attain live birth. Cloned cows often have heart and lung problems; cloned mice later develop pathological obesity; other live-born cloned animals fail to reach normal developmental milestones.

The problem, scientists suggest, may lie in the fact that an egg with a new somatic nucleus must re-program itself in a matter of minutes or hours (whereas the nucleus of an unaltered egg has been prepared over months and years). There is thus a greatly increased likelihood of error in translating the genetic instructions, leading to developmental defects some of which will show themselves only much later. (Note also that these induced abnormalities may also affect the stem cells that scientists hope to harvest from cloned embryos. Lousy embryos, lousy stem cells.) Nearly all scientists now agree that attempts to clone human beings carry massive risks of producing unhealthy, abnormal, and malformed children. What are we to do with them? Shall we just discard the ones that fall short of expectations? Considered opinion is today nearly unanimous, even among scientists: attempts at human cloning are irresponsible and unethical. We cannot ethically even get to know whether or not human cloning is feasible.

Threats to Identity

If it were successful, cloning would create serious issues of identity and individuality. The clone may experience concerns about his distinctive identity not only because he will be, in genotype and in appearance, identical to another human being, but because he may also be twin to the person who is his "father" or his "mother"—if one can still call them that. Unaccountably, people treat as innocent the homey case of intra-familial cloning—the cloning of husband or wife (or single mother). They forget about the unique dangers of mixing the twin relation with the parent-child relation. (For this situation, the relation of contemporaneous twins is no precedent; yet even this less problematic situation teaches us how difficult it is to wrest independence from the being for whom one has the most powerful affinity.) Virtually no parent is going to be able to treat a clone of himself or herself as one treats a child generated by the lottery of sex. What will happen when the adolescent clone of Mommy becomes the spitting image of the woman with whom Daddy once fell in love? In case of divorce, will Mommy still love the clone of Daddy, even though she can no longer stand the sight of Daddy himself?

> *"If it were successful, cloning would create serious issues of identity and individuality."*

Most people think about cloning from the point of view of adults choosing to clone. Almost nobody thinks about what it would be like to be the cloned child. Surely his or her new life would constantly be scrutinized in relation to that of the older version. Even in the absence of unusual parental expectations for the clone—say, to live the same life, only without its errors—the child is likely to be ever a curiosity, ever a potential source of deja vu. Unlike "normal" identical twins, a cloned individual—copied from whomever—will be saddled with a genotype that has already lived. He will not be fully a surprise to the world: people are likely always to compare his doings in life with those of his alter ego [the person from whom he was cloned], especially if he is a clone of someone gifted or famous. True, his nurture and his circumstance will be different; genotype is not exactly destiny. But one must also expect parental efforts to shape this new life after the original—or at least to view the child with the original version always firmly in mind. For why else did they clone from the star basketball player, the mathematician, or the beauty queen—or even dear old Dad—in the first place?

"Manufactured" Children

Human cloning would also represent a giant step toward the transformation of begetting into making, of procreation into manufacture (literally, "handmade"), a process that has already begun with in vitro fertilization and genetic testing of embryos. With cloning, not only is the process in hand, but the total genetic blueprint of the cloned individual is selected and determined by the human artisans. To be sure, subsequent development is still according to natural processes; and the resulting children will be recognizably human. But we would be taking a major step into making man himself simply another one of the man-made things.

How does begetting differ from making? In natural procreation, human beings come together to give existence to another being that is formed exactly as we were, by what we are—living, hence perishable, hence aspiringly erotic, hence procreative human beings. But in clonal reproduction, and in the more advanced forms of manufacture to which it will lead, we give existence to a being not by what we are but by what we intend and design.

Let me be clear. The problem is not the mere intervention of technique, and the point is not that "nature knows best." The problem is that any child whose being, character, and capacities exist owing to human design does not stand on the same plane as its makers. As with any product of our making, no matter how excellent, the artificer stands above it, not as an equal but as a superior, transcending it by his will and creative prowess. In human cloning, scientists and prospective "parents" adopt a technocratic attitude toward human children: human children become their artifacts. Such an arrangement is profoundly dehumanizing, no matter how good the product.

Procreation dehumanized into manufacture is further degraded by commodification, a virtually inescapable result of allowing baby-making to proceed under the banner of commerce. Genetic and reproductive biotechnology companies are already growth industries, but they will soon go into commercial orbit now that the Human Genome Project has been completed. "Human eggs for sale" is already a big business, masquerading under the pretense of "donation." Newspaper advertisements on elite college campuses offer up to $50,000 for an egg "donor" tall enough to play women's basketball and with SAT scores high enough for admission to Stanford; and to nobody's surprise, at such prices there are many young coeds eager to help shoppers obtain the finest babies money can buy. (The egg and womb-renting entrepreneurs shamelessly proceed on the ancient, disgusting, misogynist premise that most women will give you access to their bodies, if the price is right.) Even before the capacity for human cloning is perfected, established companies will have invested in the harvesting of eggs from ovaries obtained at autopsy or through ovarian surgery, practiced embryonic genetic alteration, and initiated the stockpiling of prospective donor tissues. Through the rental of surrogate-womb services, and through the buying and selling of tissues and embryos priced according to the merit of the donor, the commodification of nascent [unborn] human life will be unstoppable.

Changing the Parent-Child Relationship

Finally, the practice of human cloning by nuclear transfer—like other anticipated forms of genetically engineering the next generation—would enshrine and aggravate a profound misunderstanding of the meaning of having children and of the parent-child relationship. When a couple normally chooses to procreate, the partners are saying yes to the emergence of new life in its novelty—are saying yes not only to having a child, but also to having whatever child this child turns out to be. In accepting our finitude, in opening ourselves to our replacement, we tacitly confess the limits of our control.

> *"Children will hold their cloners responsible for everything, for nature as well as for nurture."*

Embracing the future by procreating means precisely that we are relinquishing our grip in the very activity of taking up our own share in what we hope will be the immortality of human life and the human species. This means that our children are not our children: they are not our property, they are not our possessions. Neither are they supposed to live our lives for us, or to live anyone's life but their own. Their genetic distinctiveness and independence are the natural foreshadowing of the deep truth that they have their own, never-before-enacted life to live. Though sprung from a past, they take an uncharted course into the future.

Much mischief is already done by parents who try to live vicariously through their children. Children are sometimes compelled to fulfill the broken dreams of unhappy parents. But whereas most parents normally have hopes for their children, cloning parents will have expectations. In cloning, such overbearing parents will have taken at the start a decisive step that contradicts the entire meaning of the open and forward-looking nature of parent-child relations. The child is given a genotype that has already lived, with full expectation that this blueprint of a past life ought to be controlling the life that is to come. A wanted child now means a child who exists precisely to fulfill parental wants. Like all the more precise eugenic manipulations that will follow in its wake, cloning is thus inherently despotic, for it seeks to make one's children after one's own image (or an image of one's choosing) and their future according to one's will.

Is this hyperbolic? Consider concretely the new realities of responsibility and guilt in the households of the cloned. No longer only the sins of the parents, but also the genetic choices of the parents, will be visited on the children—and beyond the third and fourth generation; and everyone will know who is responsible. No parent will be able to blame nature or the lottery of sex for an unhappy adolescent's big nose, dull wit, musical ineptitude, nervous disposition, or anything else that he hates about himself. Fairly or not, children will hold their cloners responsible for everything, for nature as well as for nurture. And parents, especially the better ones, will be limitlessly liable to guilt. Only the truly despotic souls will sleep the sleep of the innocent.

A Slippery Slope

The defenders of cloning are not wittingly friends of despotism. Quite the contrary. Deaf to most other considerations, they regard themselves mainly as friends of freedom: the freedom of individuals to reproduce, the freedom of scientists and inventors to discover and to devise and to foster "progress" in genetic knowledge and technique, the freedom of entrepreneurs to profit in the market. They want largescale cloning only for animals, but they wish to preserve cloning as a human option for exercising our "right to reproduce"—our right to have children, and children with "desirable genes." As some point out, under our "right to reproduce" we already practice early forms of unnatural, artificial, and extra-marital reproduction, and we already practice early forms of eugenic choice. For that reason, they argue, cloning is no big deal.

We have here a perfect example of the logic of the slippery slope. The principle of reproductive freedom currently enunciated by the proponents of cloning logically embraces the ethical acceptability of sliding all the way down: to producing children wholly in the laboratory from sperm to term (should it become feasible), and to producing children whose entire genetic makeup will be the product of parental eugenic planning and choice. If reproductive freedom means the right to have a child of one's own choosing by whatever means, then reproductive freedom knows and accepts no limits.

> **"What looks like compassionate humanitarianism is, in the end, crushing dehumanization."**

Proponents want us to believe that there are legitimate uses of cloning that can be distinguished from illegitimate uses, but by their own principles no such limits can be found. (Nor could any such limits be enforced in practice: once cloning is permitted, no one ever need discover whom one is cloning and why.) Reproductive freedom, as they understand it, is governed solely by the subjective wishes of the parents-to-be. The sentimentally appealing case of the childless married couple is, on these grounds, indistinguishable from the case of an individual (married or not) who would like to clone someone famous or talented, living or dead. And the principle here endorsed justifies not only cloning but also all future artificial attempts to create (manufacture) "better" or "perfect" babies.

Warnings of Dehumanization

The "perfect baby," of course, is the project not of the infertility doctors, but of the eugenic scientists and their supporters, who, for the time being, are content to hide behind the skirts of the partisans of reproductive freedom and compassion for the infertile. For them, the paramount right is not the so-called right to reproduce, it is what the biologist Bentley Glass called, in the 1970s, "the right of every child to be born with a sound physical and mental constitution, based on a sound genotype . . . the inalienable right to a sound heritage." But to secure this right, and to achieve the requisite quality control over new human life, human conception and gestation will need to be brought fully into the bright light of the laboratory, beneath which the child-to-be can be fertilized, nourished, pruned, weeded, watched, inspected, prodded, pinched, cajoled, injected, tested, rated, graded, approved, stamped, wrapped, sealed, and delivered. There is no other way to produce the perfect baby.

If you think that such scenarios require outside coercion or governmental tyranny, you are mistaken. Once it becomes possible, with the aid of human genomics, to produce or to select for what some regard as "better babies"— smarter, prettier, healthier, more athletic—parents will leap at the opportunity to "improve" their offspring. Indeed, not to do so will be socially regarded as a form of child neglect. Those who would ordinarily be opposed to such tinkering will be under enormous pressure to compete on behalf of their as yet unborn children—just as some now plan almost from their children's birth how to get them into Harvard. Never mind that, lacking a standard of "good" or "better," no one can really know whether any such changes will truly be improvements.

Proponents of cloning urge us to forget about the science-fiction scenarios of laboratory manufacture or multiple-copy clones, and to focus only on the sympathetic cases of infertile couples exercising their reproductive rights. But why, if the single cases are so innocent, should multiplying their performance be so off-putting? (Similarly, why do others object to people's making money from that practice if the practice itself is perfectly acceptable?) The so-called science-fiction cases—say, *Brave New World*—make vivid the meaning of what looks to us, mistakenly, to be benign. They reveal that what looks like compassionate humanitarianism is, in the end, crushing dehumanization.

Chapter 4

Should Genetic Engineering Be More Closely Regulated?

Regulating Genetic Research: An Overview

by Denis Noble

About the author: *Denis Noble is a professor of cardiovascular physiology (the study of the functioning of the heart and blood vessels) at Oxford University in England. He is co-editor of the UNESCO publication* Ethics of Life *and a founder of the Save British Science Society.*

The natural world is a very dangerous place for human beings. If evolution had taken even slightly different directions at various stages, we would not be here to contemplate the results. We may literally owe our existence to the crash of a meteorite that wiped out the dinosaurs. Another one could well wipe us out in the future.

I start with these startling thoughts because it is important to put into perspective the debate about the social and ethical dilemmas posed by recent developments in genetics and biotechnology. We exist as a result of genetic modification as we evolved, in a harsh world, from the tiny mammals that survived the demise of the dinosaurs. Out of the challenges of that harshness emerged the extraordinary intelligence that has enabled us to enter an entirely new phase of biological evolution: one driven by the rapid growth and transmission of knowledge, not just by genetic mutation.

As a result we now live in an era of unprecedented comfort and luxury for our species, and it is all too easy to forget quite how precarious our way of life is. We expect to live our full lifespans. We expect to be free of major diseases and are puzzled that some are so resistant to the attack of medical research. AIDS took us by surprise. We will, no doubt, feel "let down" by the imminent exhaustion of our armoury of antibiotics, as bacteria evolve that can resist all we have to throw at them.

In reality we are going to need all the knowledge we can glean from new research if we are to have any chance of maintaining our present lifestyles and civilisation and to chart our way through the future. So, asking for a freeze on new research—for example, on genetic modification—is a recipe for long-term decay. We must seek instead to benefit from what the new biological technologies can give. We must survive by our wits.

Unpredictable Science

There are dangers, of course, as recent public debate has highlighted. But the greater danger is to make the simplistic assumption that all forms of genetic modification are potentially dangerous. This is simply not true. There is a big difference, for example, between interventions in our genetic make-up that correct a fault in our own lifetime and those that change the germ cells, and so have consequences for future generations. A genetic therapy that counters a vulnerable individual's liability to heart attack, for example, and which is restricted to changing the expression of existing genes, is quite different from an attempt to eradicate the offending gene from future generations. Both could be options in the future. The second clearly poses more ethical dilemmas than does the first, though in the long term it may also be more effective.

Can we safely predict what the side-effects of such genetic manipulation may be? The honest answer must be that it will be difficult. It is generally wrong to think of individual genes as being genes "for" something. More often, large numbers of genes combine to produce their complex results in our bodies. A gene that has a bad effect may also take part in a combination of beneficial genetic interactions. After all, its existence suggests that something must have favoured its selection.

This is why we should insist that research on the sequencing of the human genome (identifying the molecular character of each of our 80,000 or so genes) should be complemented by research on how the genes actually function, individually and together, in whole organs and systems. Ideally this research should be a public-private partnership. The public role, funded by research councils and medical charities, should include laying down common standards to ensure compatibility of research efforts. Development and commercial application of this new knowledge will be largely the task of private companies.

Reports issued in 1999 by PriceWaterhouseCoopers on the future of the pharmaceutical industry have highlighted the enormous potential benefits to be gained from the next layer of genetic discovery. At present, for example, much research in the industry is wasted when new drugs fail because unexpected side-effects are detected in a very small minority of the population. The results of a [pounds]200 million investment can be wiped out

> *"We are going to need all the knowledge we can glean from new research."*

by a single clinical trial. But imagine a situation in which genetic typing of individuals could be so precise that one could screen out the susceptible people before they were treated. Many valuable drugs that have been banned by the regulators could have a new lease of life.

Elaborate computer modelling plays an increasingly significant role in this research. My own research team and its international collaborators have been responsible for developing a virtual heart that has already been used in drug assessment and testing. Virtual models of other organs and systems are in the pipeline.

Ownership of Knowledge

Who owns all this knowledge? The short answer is that there is a complex patchwork of ownership. Much of it is in the public domain, published in the science journals. But much is also in the private domain, protected by patents and copyright. There is probably no alternative to this rather messy situation. Exploitation of the new technologies is very costly, and the necessary capital is not going to be available without the right to patent. What is needed is sensitive regulation to ensure that the benefits, even those developed privately, are made as widely available as possible. The form of regulation should be similar to that currently in place for the pharmaceutical industry, governing the introduction of new drugs. One of the most important features of the drug-approval system is that companies consult the regulatory authorities almost from the stage of conception of a new treatment, possibly over the course of several years ahead of the final decision. This ensures that the entire process is subject to monitoring and scrutiny.

> *"Regulatory bodies will . . . need to determine what [genetic] information must stay in the public domain."*

There is also an urgent need in Europe to review the patent laws. At the moment it is impossible to patent an idea once you have published it, which produces unhelpful delays in the dissemination of new knowledge. By contrast, in the US, patenting is still possible for up to a year after publication. We should encourage this approach, as it minimises the need for secrecy and gets information out into the public arena more quickly.

Regulatory bodies will also need to determine what information must stay in the public domain. Just as the human genome sequence must be publicly available, so should the modelling of its physiological interpretation. This issue is sometimes presented as a clash between public and private needs. It doesn't need to be. What the private companies require is protection, via [by means of] patents, for their commercial exploitation of the relevant knowledge. The knowledge itself doesn't need to be secret. Nor should academic researchers be required to pay royalties for its non-commercial use.

Allied to this is a vague public unease that the generation of biotech knowledge in private companies may lead to overemphasis on exploitation for profit rather than for maximum benefit. There is an inevitable tension here between private and public needs, which it is probably too early yet to resolve satisfactorily. The biotechnology industry is still very young: very few private operations have made any profit for their investors to date; indeed, most do not even know which future commercial applications might make them money. Heavy-handed regulation of the industry at such a formative stage could easily strangle it. The regulators are not going to find it easy to develop ground rules and get the balance right.

Regulating Biotechnology

Beyond the question of ownership, there is the deeper issue of how to regulate the research itself. This is the task for governments, and preferably through international co-operation.

There are various ways in which this regulation could be given effect. Expert committees will continue to be necessary—though we have now learnt the lesson that it is important to have lay [nonscientist] representation on these bodies. It is also important to encourage the widest possible public debate: the overriding political reality is that implementation of any of these unfamiliar and unsettling scientific discoveries can only be with informed public consent. In the modern knowledge-based society people are no longer prepared to accept expert advice uncritically, even when the scientific issues are very complex. Nor should they! We must all take responsibility for the world we will hand on to succeeding generations.

The current debate in Britain on genetically modified foods demonstrates quite how important, and tricky, the politicians' job is. Faced with a highly suspicious public, the government needs to exercise considerable political leadership to put this debate back on track. The stakes are high. If we fail to do so, Britain will lose out commercially, just as it did over the exploitation of the first antibiotics half a century ago.

Finally we must recognise that scientific problems of public importance increasingly require judgements to be made about levels of risk and benefit. These cannot be value-free: we all assess risks differently according to our own likes and dislikes. That is where politics takes over from science. But the politics must be of a very different kind from what we have been used to. No longer should ministers be able to hide behind the argument that they are relying on anonymous, behind-the-scenes expert opinion. They have a responsibility both to build in a role for relevant experts within the political process, and to acquire knowledge of the science and technology themselves, so they can make informed judgements. This is not an optional challenge, but one that should be central to the government's approach as it continues its programme of democratic reform.

Genetically Engineered Foods Should Be Labeled

by Daniel S. Greenberg

About the author: *Daniel S. Greenberg is a science writer. His books include* The Politics of Pure Science *(University of Chicago Press, 1999). He frequently writes articles for the British medical journal* The Lancet.

It's difficult not to sympathize with our European cousins in their resistance to "Frankenstein foods." That's the term they apply to foods in which genes have been added, deleted or modified to make them cheaper to produce or to prolong their freshness.

The real issue is political, rather than medical or scientific. Basically, it comes down to the people's right to know what they're eating, even if their notions on that matter are demonstrably nonsensical to scientists and government regulators. In most cases, genetically modified foods are unrecognizable because the changes are hidden in the genes and the manufacturers have successfully fought off labeling requirements as invitations to hysterical boycotts.

Genetically modified foods are completely safe, according to the producers, predominantly American firms, which, after fairly smooth sailing on the domestic market, are stunned by the hostile European reaction. No reliable evidence of harm has come to light, apart from reported difficulties for Monarch butterflies in genetically modified crop fields. A scare about modified potatoes in Britain has been rejected as schlock science by an authoritative independent review.

About half of the huge U.S. soybean crop is genetically modified, and nothing Frankensteinian has yet been reported. Without evidence of harm to humans, the warnings of danger remain speculative and must be viewed against the great potential of genetic engineering for raising agricultural productivity.

A Right to Know

But the manufacturers' assurances of safety are more syrup than science. Scarcely any research was conducted on human effects prior to marketing of the products, and monitoring for possible long-term consequences is difficult and practically nonexistent.

While European supermarket chains, responding to consumer sentiment, are boycotting genetically modified foods, little sales resistance had arisen in the United States, despite a recent survey that found 81 percent of Americans favor mandatory labeling. The trans-Atlantic difference may be that Americans are accustomed to a steady stream of novel products from a highly competitive food industry, whereas Europeans tend to be more traditional about what they eat.

The U.S. acceptance may also reflect confidence in the Food and Drug Administration (FDA) and the Department of Agriculture. For all the criticism these agencies receive, the American public trusts them on food safety, which is not the case with their many European counterparts. After a flurry of negative publicity, bovine growth hormone, which raises cows' milk production, was approved by the FDA in 1993 and has since attracted little notice in the United States, despite extensive use. But its use has been banned in most European countries and elsewhere, too. With the milk hormone, as well as with other genetically modified products, the FDA says that mandatory labeling is not required because genetic modifications do not significantly change the food.

> *"The people have a right to be informed and to arrive at the wrong conclusion."*

Essentially the regulators are telling us that, as far as science can tell, it's all in the head of the consumer. Therefore, why burden manufacturers with a labeling requirement that can discourage sales? The policy reflects deep respect for science, and it calms the fears of the biotechnology industry, which acknowledges generous government support for research—with generous campaign contributions to both political parties. Nonetheless, the no-labeling policy misuses the authority of science in trampling over the right to know what you're eating.

For religious reasons, many Jews and Muslims refuse to eat pork. To guide them in their preference, packaged products indicate the presence of pork, though there is no scientific evidence that it is unwholesome. Package labels also assist vegetarians, as well as persons who are sensitive to particular foods, or who think they are.

Scientists are probably right when they insist there's nothing to worry about in genetically modified foods. But in this matter, as well as others, the people have a right to be informed and to arrive at the wrong conclusion. Science can serve as a guide, but it shouldn't perform as a bully.

Patents on Genetically Engineered Crops Should Be Limited

by C.S. Prakash

About the author: *C.S. Prakash teaches plant molecular genetics and is director of the Center for Plant Biotechnology Research at Tuskegee University. He is a member of the USDA Advisory Committee on Agricultural Biotechnology. He is also the president of the AgBioWorld Foundation, a nonprofit organization that provides information about and supports biotechnology in agriculture.*

Bioengineered crops were grown on nearly 40 million hectares (100 million acres) in twelve countries last year—up from less than two million hectares when they were first introduced in 1996, making biotechnology the most rapidly adopted technology in the history of agriculture. But this phenomenal success has been a double-edged sword. Despite the certified safety of biotechnology-derived foods, opposition by environmental activists has undermined consumer confidence in the new gene technology. Food companies such as McDonald's and Frito-Lay are now asking their suppliers not to use bioengineered potatoes and corn. Many European countries are avoiding imports of bioengineered corn and soybeans entirely.

Meanwhile, the industry has responded with a public relations campaign of its own. The press releases and TV commercials extol potential benefits of biofoods, such as better nutrition and ameliorating the problem of world hunger. Although biotechnology clearly provides ammunition for improving food production, the fact is that right now there is little industry research on food staples of importance to the developing world. It's time for the industry to put its money—actually its patents—where its mouth is.

A Barrier to the Hungry

Nobody should expect Monsanto to end world hunger. That's like counting on Microsoft to wipe out illiteracy. The biotech industry has spent billions of dollars developing a powerful technology for redesigning crops to evade pests and diseases, and to improve food quality. But because investment dollars need to be recovered, the target of such research is on commercial crops in Western countries.

So where does that leave the developing world? Poor countries such as Ethiopia or Bangladesh don't have the funds or scientific talent needed to pursue biotech research on their own. Nevertheless, many public institutions are developing food crops with improved attributes such as "golden rice" rich in provitamin A, which can prevent blindness in children. In my own lab at Tuskegee University, we have created high protein sweet potatoes.

These new crops are designed to be distributed freely to farmers in the developing world. However, industry "ownership" of genes and technologies used to create such varieties represents a serious obstacle. Nearly every core technology used in crop biotechnology is the intellectual property of companies such as Dow, DuPont, Monsanto and Novartis. So if Vietnam or Liberia wants to distribute golden rice seeds to its farmers, it must first negotiate with various companies for the gene transfer, gene promoter and selectable marker technologies that were used in its development. Most poor countries simply do not have the financial resources or the scientific or legal acumen [knowledge and skill] to wade through this complex patent maze. Thus, agricultural biotechnology cannot make inroads into developing nations without a "freedom to operate" license from the owners of these technologies—major life science corporations.

> *"If companies really want to combat global poverty and hunger, they must make their technology available for use . . . by developing countries on a royalty-free basis."*

A Call for Sharing

If companies really want to combat global poverty and hunger, they must make their technology available for use on select food crops such as rice, cassava and millet by developing countries on a royalty-free basis. Not only will this provide a tremendous boost to world food production, but it also makes good business sense. Acceptance of biotech food crops in the developing world would create market opportunities for commercial crops such as cotton, and would also give the industry a much-needed human face.

Would anyone oppose such a plan? Although there's much willingness among corporate scientists to share technology, their lawyers cannot see beyond the issue of liability. Activists are also to blame. Their opposition to using new technologies in the Third World puts industry in a "damned if you do and damned if you don't" position.

Clearly, we need an independent middleman to take charge. Catherine Ives of the Agricultural Biotechnology Sustainability Project at Michigan State University believes that a new international agency should be set up to act as a "technology trust" that can assume responsibility for transferring biotechnology to developing countries. A central agency would not only help indemnify companies from liability suits, but would also help negotiate the labyrinth of patent laws and intellectual property claims.

The benefits of agricultural biotechnology are as real as the problems we face. In my native India, every third child is underweight due to malnutrition and 400 million people go to bed hungry every night. In a country where 70 percent of people are associated with farming, technological innovation in agriculture is critical not only to produce more food but also to improve living standards. It's time for the agricultural biotechnology industry to show a social conscience and clear the way for the harnessing of their newfound knowledge to combat global hunger and malnutrition.

Patents on Human Genes Should Be Limited

by Seth Shulman

About the author: *Seth Shulman frequently writes about intellectual property issues. His books include* Owning the Future *(Houghton Mifflin, 1999).*

Who owns the human genome? A profound confusion reigns today over the issue of proprietary rights to human genes, and it is setting the stage for a long-term intellectual-property disaster. The U.S. Patent and Trademark Office, backlogged with tens of thousands of private claims to genetic information, cannot seem to decide where to draw the line on ownership rights; animosity between government-funded genome researchers and their industry counterparts continues to mount; and an ominous tangle of lawsuits already looms on the horizon.

And this, as everyone in the field will tell you, is only the beginning of many decades of fast-paced genomic research.

The controversy was simmering just behind the happy facade presented during the White House ceremony in June 2000 when President Clinton announced the completion of two working drafts of a human genome map. Despite the grand announcement, there is a startling uncertainty about how to apportion the lucrative information the map contains. For an economic indication of this uncertainty, look no further than the joint statement Clinton and British Prime Minister Tony Blair issued in March 2000. In the statement, Clinton and Blair did nothing more than affirm the widely agreed-upon need to maintain open access to the human genome's raw genetic sequences. Then they watched dumbfounded as a stunning flight of capital bled billions of dollars from the biotech companies involved in the genome field. Some firms, such as Palo Alto, Calif.–based Incyte Genomics, lost nearly one-third of their stock market value in one day.

The irony in Wall Street's jittery response is that firms like Incyte, Millennium Pharmaceuticals, Celera Genomics and Human Genome Sciences—the most vociferous proponents of proprietary rights over human genes—all clam-

From "Toward Sharing the Genome," by Seth Shulman, *Technology Review*, September 2000.

ored to explain that they welcomed the Clinton-Blair announcement. Why? Because these companies recognize that some level of open access to the human genome is essential to building a robust industry in the future.

A Patenting Gold Rush

These genomics firms see the tremendous potential of the decoded human genome to medical science. They know, in more detail than most of us, that the diagnostic screening tests and drugs now coming to market offer only a glimmer of the promise ahead. They also understand something else: They have the most to lose if they are shut out from developing these treatments by their competitors' capricious and overly broad intellectual-property claims.

The race to discover and patent human genes is frequently likened to a gold rush, an analogy that captures much of the current frontier flavor. But the savviest players recognize a fundamental difference. Prospectors forever removed the gold when they panned it from rivers and mined it from the earth. But the information in the human genome is not depleted upon its discovery. The genome is a resource to which biomedical researchers will return again and again to solve the puzzles of human disease; it is a wellspring that will nurture a myriad of overlapping discoveries and inventions for many decades to come. In this sense, the human genome can—and must—support what economists call nonrival consumption: a situation where multiple parties profitably share the same resource.

The problem with the gold rush analogy, then, is that the land claims that helped tame the prospectors' free-for-all were a crude but necessary framework to divide rights to a tangible and decidedly finite resource. By carving up the genome into parcels of exclusive, private real estate, the Patent Office is needlessly replaying this history. Instead, what is called for is an enlightened policy designed to govern the multiple and overlapping uses of the "genome commons": a policy that insures unfettered access to the data and materials that will serve as the building blocks for countless drugs and treatments in the future.

A "Genomic Commons"

Although the term "commons" traditionally refers to shared property such as a city plaza or communal pastureland, the notion also applies to shared knowledge resources. We would do well to remember the 1799 discovery of the Rosetta stone, the remarkable tablet that offered the same long passage of text sequentially in three ancient languages. The Rosetta stone provided linguists over many ensuing years the seminal [basic] key to finally unlock the previously undecipherable hieroglyphics of ancient Egypt. Imagine if someone had proposed to chop the stone tablet into separate proprietary chunks. Such a plan would clearly have diminished, if not destroyed, the central value of the resource.

Sadly though, that is exactly what is happening now with the human genome. Companies must think about their financial bottom line. For today's crop of genomics firms, especially given the absence of clear rules, this means obtaining patents—as many as possible and with the broadest possible claims. The tens of thousands of human gene-related patents pending have polarized an already divisive situation. On the one hand, companies investing millions of research dollars argue that they need to protect their intellectual property. Without patents, they say, the private sector won't ante up the billions of dollars needed to stimulate the rapid development of genome-based healthcare products. On the other hand, the patenting frenzy is kindling understandable fears that only a few corporations will end up controlling a resource of priceless value to humanity. Some, like biologist Jonathan King at MIT, who is circulating a petition to this effect, believe the answer is to prohibit patents on genes.

Between these views is a gray area as big as the Pacific Ocean. As University of Michigan legal scholars Michael Heller and Rebecca Eisenberg explain it, the key question is how far "upstream" or "downstream" proprietary rights ought to be allowed along the path to product development. The problem is that almost any piece of the genome can be seen to have some commercial value. But issuing patents too far upstream could leave the path to drug development looking like a pockmarked road with a nightmare of tollbooths and barricades.

> *"Some level of open access to the human genome is essential to building a robust industry in the future."*

While the specifics can admittedly be confusing, at root there is nothing particularly complicated about the debate over patenting and the human genome. The key is to treat this as a vital public policy issue rather than a strictly legal or scientific one. The first thing to remember is that the human genome is a precious inheritance of the human species. For this reason alone, it deserves special treatment. Second, the project to decode the human genome has, for more than a decade, been the mission of a publicly funded project that will ultimately cost some $3 billion. Given this outlay of funds by taxpayers, the public has every right to demand that the genome is used wisely and not simply handed over for private gain. And finally, we must recognize that—with the recent, momentous milestone of a completed working draft of the human genome—the time to act is now. We can save ourselves a lot of bitter litigation, money and acrimony (and speed the next generation of drugs and treatments to market) if we adopt policies that balance the powerful drive of commercial interests with the public interest.

Here are five steps to take immediately:

1. Create the World's First "IP-Free Zone"

The publicly funded Human Genome Project began in 1988 as a grand scientific undertaking to decode all human genes and create a shared resource of knowledge that will be the foundation of 21st-century medicine. It's time to codify that status in law by creating the world's first intellectual-property sanctuary, or "IP-free zone." Specifically, the U.S. government should mandate that the genome's raw sequence data cannot be privately owned.

In a practical sense, we are already well on our way to this goal. Over the past several years, the Human Genome Project's network of laboratories have published on the GenBank Web site (www.ncbi.nlm.nih.gov/Genbank/) a steady stream of decoded nucleotide base pairs within 24 hours of sequencing them. The policy insures that the genome data will be freely

> *"The human genome is a precious inheritance of the human species."*

available to researchers around the world, and also discourages secrecy or proprietary claims over this valuable raw data.

With the Human Genome Project's laudable commitment to open publication (thereby de facto [in effect] disqualifying the raw data from patent claims), why bother writing a law? Because the formal establishment of an IP-free zone—a kind of "national park" of genome knowledge—will set a vital precedent that some kinds of precious information resources must be off-limits for private ownership.

As many are coming to realize as we move headlong into the knowledge-based economy, things work best when seminal information assets—particularly those needed by all players in a given high-tech sector to compete—are pooled and shared. This is the idea behind both open-source software and the hardware standards that have come to predominate in many high-tech sectors. These pieces of "infostructure" are what allow different engineers to design distinct machines that all plug into a single type of wall socket or send standardized software files over the Internet. Like public lands or public libraries, pooled knowledge assets must be made freely available and protected within a framework that preserves their integrity. Precisely because the raw sequence of the human genome—the actual string of DNA base pairs themselves—is already widely perceived as deserving of this special status as a category of knowledge, it makes an ideal candidate for the world's first legally mandated IP-free zone.

2. Declare a Moratorium on Gene Patenting

The U.S. Patent and Trademark Office has clearly bungled the issue of gene patenting by years of equivocation and delay. We need a moratorium on gene patenting until we can all agree on sensible rules.

Patents have always represented a compact between an inventor and the public: The inventor gets a 20-year monopoly in order for the public to quickly get the benefit of the new innovation in the marketplace. Issuing broad patents that confer no such clear benefits can, as the Supreme Court noted in the key 1966 case *Brenner v. Manson*, create a "monopoly of knowledge . . . [that] may confer power to block off whole areas of scientific development, without compensating benefit to the public."

The most urgent task is to clarify that before a gene can be patented, it must be shown capable of providing a benefit to the public—namely by bringing a new product or invention to the marketplace. Within our legal framework, this principle is defined as a patent's "utility."

In a hopeful sign, the director of the Patent Office, Q. Todd Dickinson, has proposed to raise the "utility" bar slightly by adding three words to its guidelines. To win a patent in the human genome field, applicants will have to describe a "substantial, specific and credible" use for their gene. While this represents an important step in the right direction, the Patent Office can and should do more. Although Dickinson may rightly argue that his is not a policymaking body, the Patent Office's role on the front lines requires it to spot emerging problems and report them to Congress or otherwise work to redress them. To stem the continuing confusion in this area, Dickinson needs to take the lead, holding hearings that include all stakeholders and declaring a moratorium on gene patenting until acceptable rules are reached.

3. Institute a Licensing System

The most troubling aspect of the current spate of gene patenting is the risks that the right afforded by patents "to exclude others" poses to medicine. To avoid the worst potential problems, we need a compulsory licensing system, and Congress should hold hearings as soon as possible to figure out how it should be implemented.

Changing licensing rules would go far to insure that in the future researchers will be guaranteed the right not just to view the genome but also to use the information it contains to develop vital products. Under a compulsory licensing plan, patent holders could still receive compensation, but would no longer retain the right to exclude others from research that could improve public health. Congress needs to spell out new rules in this regard, specifying guidelines for licensing fees and perhaps adopting a sliding scale of payments for public sector vs. commercial research.

Without new licensing rules we risk replaying the sorry history of antitrust actions and patent tangles. A century ago, broad patents tied both the automotive and aviation industries in knots during their first decades. The costly and debilitating tangle of lawsuits only ended with the creation of so-called patent pools that cross-licensed

> *"The U.S. government should mandate that the [human] genome's raw sequence data cannot be privately owned."*

everything and divvied up the royalties. The patent pools—later used in the semiconductor industry—can successfully end the worst of the wrangling. But they also tend to consolidate entrenched oligopolies [concentrations of power in the hands of a few companies] and present stiff barriers to entry in the field.

If at all possible, in the case of the human genome we need to try to avoid acrimonious after-the-fact kinds of settlements. A new, transparent licensing system will help companies navigate what's likely to be an unusually complex terrain of patent claims and ensure they have access to what they need to create new drugs.

4. Establish a Zoning Commission

The management of the human genome is a vital matter of public policy. It is not something that should be left to powerful business players, lawyers or scientists to resolve alone. We need a body made up of stakeholders from the public and private sectors, scientists and laypeople [nonscientists] alike, to serve as a kind of genomic zoning board.

In the first decade of recombinant DNA research, the National Institutes of Health (NIH) established a body called the Recombinant DNA Advisory Committee (known as "the Rack") to determine policy about what safeguards were needed for certain kinds of previously untried recombinant DNA research. Although the practice of gene patenting also presents many unknowns, there has been little analysis of its economic, social or scientific impact. There isn't even an official body within the NIH to deal with the issue. It is naive to think that a laissez-faire policy [a policy with few restrictions] can be effective. We badly need a standard-setting group that can help shape policy in this emerging area.

> *"The most important thing is to remember the overriding public health mission of human genome research."*

Why think of it as a "zoning board?" Because zoning regulations represent a good example of how community standards can be established and enforced to shape the rules governing private property. We would do well to remember the lesson of Central Park in New York City. At the time it was proposed, the idea of setting aside 700 acres of prime real estate in one of the world's most vibrant cities was controversial to say the least. The land seemed too valuable to turn into a park, but time has shown that, for most Manhattan residents, the value of Central Park far exceeds the monetary worth of the property.

Just as zoning can add value to a community by balancing the needs of the private and public sectors, we need a public policy body tasked with balancing and nurturing the partnerships between the private and public sectors that have brought us to this stage in our understanding of the human genome.

5. Put Public Health First

Ultimately, the most important thing is to remember the overriding public health mission of human genome research—public or private. While it is tempting to try to speed development with patent incentives to private firms and universities, the impulse must be balanced against longer-term concerns for how the ownership rights to genes will ultimately affect the equitable dissemination of health-related products.

The human genome is already raising vexing policy questions about our rights to confidentiality for our genetic health information and protection against genetic discrimination by insurance companies and in the workplace. Just like gene patenting, these kinds of emerging issues will require new solutions and proactive public policy. The time to start is now. As we begin the process of balancing commercial interests and public health, we need to be guided by our sense of fairness, by our democratic values and by established methods of transparency and reporting that will insure our ability to publicly regulate the situation in the future.

President Clinton's joint announcement with Prime Minister Blair in March barely scraped the surface of the vital issue of ownership of the genome and unfortunately did little to stem the continuing uncertainty in the field. But, rhetorically at least, Clinton got it right when he called upon us to "ensure the profits of human genome research are measured not in dollars but in the betterment of human life."

Genetically Engineered Foods Should Not Be Labeled

by Douglas Powell

About the author: *Douglas Powell is an assistant professor in the department of plant agriculture at the University of Guelph in Ontario, Canada. He is also scientific director of the Centre for Safe Food. He has written several articles about agricultural biotechnology.*

In 2000, one of my farmer colleagues grew some genetically engineered sweet corn and table potatoes. Neither the Bt sweet corn nor the potatoes required any insecticides to manage pests. After harvest, the two crops were sold in his farm market in Hillsburgh, Ontario (Canada), fully labelled, alongside their conventional counterparts.

The genetically engineered Bt sweet corn outsold the conventional by a margin of 3-2. Same for the potatoes. The two products were sold for the same price, and while many consumers were more interested in taste, for others, the primary selling point was the reduction in pesticide sprays and worm damage.

So why not just label all GMOs (genetically modified organisms), as such foods are routinely, though mistakenly, called?

When several small suppliers of organic foods were asked by Loblaws [Canada's largest food distributor] to remove the GMO-free claims from labels—in some cases using a black felt pen—the latest round of simplistic declarations about consumer choice was underway. But as the Canadian General Standards Board has discovered in its belaboured attempt to devise meaningful standards for GMO labelling—and as several countries attempting to implement GMO labelling have learned—the task is complicated by definitions, trace amounts and, primarily, politics.

GMO Labels Are Unnecessary

Canada already has mandatory labelling for any novel food that differs in nutritional, chemical or toxicological characteristics from its conventional counterpart, and for any food that involves the transfer of genes from known allergens. But should other GMOs be labelled? The answer is no, for the following five reasons:

1. Whole foods are not trace ingredients. Sweet corn and table potatoes are the only whole, genetically engineered foods available to be grown in Canada. It is in the commodities—field corn, soy and canola—that genetic engineering is more common. Field corn products such as high fructose corn syrup and sorbitol, for example, are ubiquitous in soft drinks and toothpaste, respectively.

What of the potato, now engineered to resist the Colorado potato beetle, the scourge of North American potato growers, with the Bt-toxin from the bacterium *Bacillus thuringiensis?* Should the French fries sold at the local ballpark—possibly derived from Bt-containing potatoes and cooked in oil derived from genetically engineered canola—carry a sticker proclaiming the involvement of biotechnology? If there is no health risk, as Health Canada has decided in the case of both of these products, the answer is no.

Should consumers be able to determine the origins of the fries? Absolutely. But that can be accomplished through 1-800 numbers, point-of-sale information and other vehicles. Mandatory labelling of all food ingredients is expensive and, in a society that values convenience over cooking skills, impractical.

2. Surveys lie. But don't some 90% of consumers want GMO ingredients labelled? Sure, but that is largely because of the way survey questions are asked. When Americans were asked what information they would like to see added to food labels, without reading a list of current triggers such as genetic engineering, over 80% failed to suggest anything.

GMO Labels Can Be Misleading

3. GMO-free often isn't. As reported in the *Wall Street Journal* in April 2001, supposedly GMO-free Yves Veggie Cuisine from Canada, along with dozens of other such products, was found to contain significant quantities of genetically engineered corn and soy. Of the 20 so-called GMO-free organic products tested for the *Journal,* 11 were found to contain ingredients from genetically engineered crops. So Loblaws is justified in demanding a verifiable standard for GMO-free from its suppliers who want to make that claim.

"GMO [genetically modified organism]-free often isn't."

4. GMO cannot be defined. The vast majority of foods are genetically modified. (GMO-free proponents are likely referring to genetically engineered foods). But that won't stop hucksters from trying to make a buck at the grocery store, or the beer store. Millions of health-conscious Canadians were undoubtedly relieved to learn in early June

2001 that Quebec's Unibroue beers contained no GMOs, joining GMO-free Canadian Club whiskey and McCain French fries in a breakfast of champions.

But the April 2001 newsletter of the Association Generale des Producteurs de Mais, which represents France's corn growers, states that biotech genes are found widely in French-grown corn. And Unibroue says nothing about usage of European malt made from European barley varieties—almost all of which contain mutated genes created by deliberate exposure to high levels of nuclear radiation technology, which is termed genetic modification under Canada's Food and Drugs Act.

> *"Labelling is not about choice."*

The GMO-free scam is also hugely susceptible to fraud and to use as a non-tariff trade barrier. Some commentators have already suggested that Loblaws enforced the no-GMO-free labelling policy to bolster the prospects of its own organic line. China's new rules on such labelling are already being interpreted as a pre-emptive trade barrier to protect its grain farmers after it enters into the World Trade Organization.

GMO Labels Limit Consumer Choice

5. Labelling is not about choice. Greenpeace and other activist groups state plainly in their literature that the products of genetic engineering may cause some unknown, theoretical health or environmental harm and should therefore be banned. However, in the absence of a ban, everything should be labelled to provide consumer choice—and that will produce a de facto ban [the same effect as a legal ban].

The number one selling tomato paste in the U.K. [United Kingdom, or Britain] in 1998 was made by Zeneca, sold in supermarkets at a slightly lower price, and labelled as derived from genetically modified tomatoes. But when a media frenzy arose in the U.K. in the fall of 1998, stores rushed to remove genetically engineered products, including the tomato paste. So the previous number one seller was no longer available. And still isn't.

Couldn't happen here? When two local Zehrs supermarkets asked us to provide genetically engineered sweet corn to their stores last fall, they were overruled by corporate headquarters in Toronto. Why? Too much controversy. Yet we had shown that consumers preferred the genetically engineered product. For some, pesticide reduction is intrinsically valuable, but to most, it becomes valuable when it is linked to an increase in quality. Labelling a whole food like sweet corn is vastly different from labelling an ingredient in a processed food, especially when the label is designed to alarm rather than inform. The biotechnology debate to date has involved superficial stereotypes, caricatures and the mindless banter of pro versus anti. It misses the point that providing food consumers are actually interested in buying involves a series of trade-offs and considerations that are specific to individual farms and locales.

Patents on Genetically Engineered Crops Need Not Be Limited

by David L. Richer

About the author: *David L. Richer is connected with AgrEvo UK Limited, a British agricultural biotechnology company now owned by Aventis.*

> Everything that can be invented has been invented.
>
> CH Duell, US Patent Office, 1899

Invention did not cease at the end of the last century, and as is all too apparent, the world is faced with an unprecedented explosion in technology. Not all of this is universally welcomed—the irritation of mobile telephones on public transport comes immediately to mind—but the new technologies affect every area of our lives. Nowhere is this more true than in agriculture.

Changes in farming proceeded slowly but steadily for thousands of years, but accelerated during the last two centuries as scientists and other observers came to understand the farming process, the need for particular nutrients and rotations [alternation of crops], and the nature of pests and diseases. Further acceleration followed after the 1950s with increased mechanisation and the introduction of effective pesticides, herbicides and fungicides. Now, in the last decade of this century, even in the last half of the 1990s, biotechnology is accelerating the pace of change once again.

Biotechnology provides a major opportunity for meeting the nutritional needs of an expanding world population with a finite resource of land. It offers a new approach to the control of pests and diseases; it will provide crops of improved nutritional quality; and it will bring increased yields.

Biotechnology is unlikely to be a complete solution to our agricultural problems, at least for the foreseeable future, but it will play a key role in a sustainable agriculture which also uses integrated pest management and plant breeding techniques. That is the expectation in developed countries, but it is only a hope for many others. There is a pressing need for the agricultural revolution to spread throughout the world, but to achieve it, we must provide an incentive to the innovators and owners of the new crop production technologies to share them.

There are many forms of encouragement—from argument to finance—and one of these is the subject of this paper. The process of technology exchange will be encouraged and facilitated by a strengthening of intellectual property laws, especially those of the developing countries. Unfortunately, like biotechnology, intellectual property rights are controversial and often misunderstood.

I propose to address some of these misunderstandings, and to indicate how strengthening intellectual property rights will enable farmers throughout the world to receive the latest developments in crop production.

What Is Intellectual Property?

Intellectual property is a broad term used to cover patents, designs, trade marks, plant breeders' rights, copyright and trade secrets. All of these have a part to play in the development and commercialisation of plant production products. However, in the context of agriculture, the three most important intellectual property rights are patents, plant breeders' rights and trade secrets. None of these creates as much argument as patents.

A patent is a monopoly of limited scope, granted to the owner of an invention, for a limited period of up to 20 years. It is a right which is effective *only* in the country which grants the patent. While it is in force, a patent enables the owner to exclude others from using the invention commercially in that country.

A patent provides the innovator with a limited period within which he has the opportunity to recoup his investment in the research and development of the invention. In return, the inventor discloses the invention to the public, and that disclosure enables other scientists and interested parties to use the invention in their own research. In due course, that research may lead to further innovation, and society will benefit. It is no coincidence that those countries with strong research-based industries are also those countries with strong intellectual property laws.

TRIPs: An International Standard

Intellectual property rights are national in character, and like other national laws they vary from country to country. Attempts to bring some harmonisation into this area have succeeded in the Agreement on Trade Related Aspects of Intellectual Property Rights, generally referred to as TRIPs. TRIPs, which entered into force in 1995, applies to all members of the World Trade Organisation (WTO). However, TRIPs is not a complete remedy for inadequate laws since it lays down only minimum levels of protection, rather than providing for the optimum. Nevertheless these levels are important, none more so than the basis for a patent set out in Article 27.

> Article 27 provides that patents shall be available for any inventions, whether products or processes, in all fields of technology, provided that they are new,

involve an inventive step, and are capable of industrial application.

Patent rights are of little use if they cannot be enforced, and TRIPs also provides that member countries shall ensure that enforcement procedures are available under national law so as to permit effective action against any act of infringement of an intellectual property right. Enforcement is a particular concern in the field of biotechnology where the capability of biological materials carrying genetic information to self-reproduce makes the copying of an invention and the infringement of patent rights all too easy.

The implementation of TRIPs will undoubtedly strengthen intellectual property rights in many parts of the world, but implementation is unlikely in the short term. Developing countries are permitted a transitional period, until 2005, within which to bring their intellectual property laws into compliance with the minimum standards laid down in TRIPs, but unfortunately, many developing countries lack the means rather than the will to take the necessary steps.

Purposes of Intellectual Property Laws

The objectives of intellectual property law are stated succinctly in TRIPs:

> The protection and enforcement of intellectual property rights should contribute to the promotion of technological innovation and to the transfer and dissemination [spread] of technology. . . .

It is useful to consider these two aspects separately.

The costs of developing a new plant protection chemical are over $150 million; the costs of developing a new transgenic plant are comparable. The investment in research and development must be recovered, and the monopoly period—provided firstly by patents and secondly by protection of the confidential data supplied in regulatory packages—is essential to provide the innovator with sufficient time and opportunity to make that recovery. Without the intellectual property protection, research-based companies would be unable to bear the risk of the major investment in research and development required to bring those technologies to the market.

The incentive effect of patents for developing countries are sometimes questioned on the grounds that these countries have little private sector research and are likely to produce few inventions. It is certainly true that inventors in those countries file few patents domestically or abroad, but without adequate intellectual property rights, there is little incentive for

> *"The process of technology exchange will be encouraged and facilitated by a strengthening of intellectual property laws."*

local companies to set up their own research departments, nor for foreign companies to bring their technology and their research capabilities to the countries. It is left to the public sector to be the major fund provider of this research. That funding is vital, but it is not sufficient.

Technology Transfer

Farmers must have the opportunity to obtain modern crop production products at a reasonable price. Local research organisations need access to the latest technologies in the form of transfer of materials and know-how to further their own research, research which is often necessary to provide solutions to peculiarly local problems. These requirements can only be met in developing countries by means of technology transfer.

The importance and benefits of technology transfer are widely recognised, and it is consequently a cause for concern and regret that in the field of crop production, technology transfer proceeds so slowly. While part of the problem lies in funding, another factor results from the two sides—the technology providers and the technology receivers—viewing each other and the technology transfer process with suspicion.

The private sector, potentially the major provider of new technology, ought to be eager to provide technology which will lead to the development of new markets. However, companies are worried that providing their know-how, whether by sales or by licensing, is tantamount to giving the technology away unless it has the protection of enforceable intellectual property rights. In the case of biotechnology, where reproduction of plant material is relatively simple, companies may be powerless to prevent their technology from being copied and their markets destroyed or undermined by those who have not incurred the expenses of developing the technology.

> *"Without intellectual property protection, research-based companies would be unable to bear the risk of the major investment in research and development."*

On the other hand, the developing countries which want and need the technology, fear the technology-provider's demands for stronger intellectual property rights which, they believe, will lead to higher prices and a drain on currency reserves. These concerns are real—and are discussed later—but there is clear evidence that strengthening intellectual property laws leads to an increase in technology transfer to the benefit of both the provider and the recipient.

To take one example, in the recent past, companies in developed countries were reluctant to bring their products into the Chinese markets. There was inadequate patent and data protection, and once companies had established their market, generic manufacturers rapidly appeared to reap the benefit. Worse, these local manufacturers exported the products to neighbouring states where there was either weak or no patent protection, or where enforcing patents required a long and uncertain litigation process. As a consequence, China was deprived of the latest plant production products.

China has recently strengthened its patent law, and although it still needs improvement in its enforcement procedures, companies from the developing world are now not only prepared to collaborate with Chinese companies, they are actively seeking collaboration.

Concerns About Intellectual Property

There are many concerns surrounding intellectual property law, and although these are often based on misunderstanding, they remain a barrier to progress. It is essential that those of us who believe that strengthening intellectual property rights will be beneficial, should listen and inform.

As a contribution to this process, I would like to consider four specific issues:
• Prices
• Local development of technology
• Theft of Resources
• Morality
This is by no means an exhaustive list of possible topics.

Prices

It is argued that intellectual property rights lead to an increase in prices. While it is true that products containing new technology will generally be sold at higher prices, that is not the same thing as a general increase in prices. The old technologies remain; indeed the introduction of new technologies may well make the old ones cheaper.

Genetically engineered seeds are more costly to develop and produce, and those increased costs must be recovered. Farmers expect to pay a higher price for seed which will bring added value, but no farmer will pay more for the benefit than the increase in value which it will provide. Producers have little choice but to price their products so as to share the added value with the farmer.

Such arguments will be of little interest to poor and subsistence farmers. In order to receive the benefits of the new technology, they must first acquire the seed. Intellectual property will not be of much help to them. They will require the assistance and support of Government agencies and organisations such as the World Bank.

It is also argued that without intellectual property rights, local companies would be able to copy the products and bring them to the local markets at much cheaper prices. That may be true but the advantages—such as they are—are short term, and serve to delay the introduction of new products.

> *"The introduction of new technologies may well make the old ones cheaper."*

Moreover, if the originator ceases to act as a product steward for the products, the result is often a flow of sub-standard products with inadequate instructions for their use.

The risk of sub-standard products should not be under-estimated. In a country, which for legal reasons I cannot name, my company is presently trying to stop the sale of low cost counterfeits of some of our products. For the farmer, the availability of low cost alternatives appears to be only to his advantage. In fact, the counterfeit products are almost entirely water with just a trace of pesticide which happened to have remained in the previously-used container.

Local Development of Technology

Some opponents of intellectual property claim that patents inhibit local research, and interfere with the work of local companies and research organisations.

The freedom to carry out research is safeguarded under patent law; experimental use for research purposes is not an infringement of the rights of the patent owner. Scientists are free to take the invention, to modify it, and to incorporate it into their own research programmes. Dissemination and use of knowledge in this way is a fundamental part of the original contract between the owner of the patent and the state granting the right. It results in the faster development of the technology, and the introduction of new products and processes.

Public funding of research in developing countries plays an essential role in addressing local problems, but it will never be sufficient. There are always other demands on the available money. Local companies and research organisations need inward investment, if not in cash then in the form of materials and know-how. Yet again, however, it has to be acknowledged that these materials and know-how have a value to the private sector which will be unwilling to supply them if it feels that in doing so, it will lose control over them.

Publicly-funded research organisations themselves do not always make the best use of the intellectual property protection which is available. The International Agricultural Research Centres, for example, have tended to favour not seeking intellectual property protection, a position which was remarked on in the 1996 OECD Survey "Intellectual Property Technology Transfer and Genetic Resources":

> [The] Centres have to operate in a changed research and funding environment and to collaborate with organisations for which intellectual property is a necessary counterpart to their willingness to invest in development. This has long been true of industrial organisations, and academic and public sector organisations are also now taking a more positive attitude towards protecting innovations resulting from their research. The International Agricultural Research Centres may therefore wish to review their own positions in this respect.

Patenting the results of their research will not prevent the Research Centres from making them available, but it will give them the option of entering into cross-licensing agreements or collaborations with companies holding other intellectual property of interest. Patents become bargaining chips which can be traded to further the research aims of the Research Centres.

A recent attack on the private sector is that companies in developed countries are stealing the resources and know-how of local populations, patenting these resources, and then denying the use of the technology to the population who had used it, often for centuries. The case of the Indian neem tree is often quoted.

Indian farmers have used the seeds of the neem tree for pest control for centuries. The American company, W R Grace, discovered a process for extracting the oil from the seeds, and applied for patents. Alarmists spread the story that the Grace patents would prevent farmers from continuing to use their traditional methods of pest control. The story created understandable consternation amongst Indian farmers and a world-wide outcry against big business. The story is nonsense.

Patents are granted only for inventions which are new and not obvious, and the use of the neem seeds in pest control fell into neither of these categories when Grace applied for its patent. Grace could not monopolise the use, and nor could a patent give Grace ownership of the neem tree or its seeds (Grace buys seed on the open market). And finally, no patent can stop anyone from doing something which he was doing before the patent application was filed. Similar stories are now circulating concerning the so-called theft of genes by the developed world. If this does occur, then it will be illegal under the provisions of the Convention on Biological Diversity.

> *"No patent can stop anyone from doing something which he was doing before the patent application was filed."*

In any event, it is worth stressing again that intellectual property rights extend only to the new inventions created from the isolation and transference of the gene. The identification of a gene with a useful trait in a local plant, and the transference of that gene into a different crop plant may entitle the discoverer to patent the use of that gene in the transformed plant, and perhaps to the transformed plant itself. However, the patent will give the innovator no rights over the original plants, which can continue to grow or be grown without reference to the patentee.

Morality

TRIPs provides that inventions may be excluded from patentability if their exploitation should be prevented in order to protect ordre public [public order] or morality. It is important to note that it is the exploitation of the invention which is concerned here, not the invention itself. This distinction has been missed by many, and has resulted in the morality arguments being extended from the use of biotechnology to biotechnology itself, and from there to biotechnological patents. Parties who believe that biotechnology is immoral also argue that patenting biotechnological inventions—or, more emotively, patenting life—is immoral. Many patents, particularly in Europe, are presently being attacked on these grounds.

The consequences of attacking patents on moral grounds may not lead to the results which opponents of biotechnology hope, a point noted by Professor

Richard Jefferson in his Expert Paper prepared for one of the Secretariats advising on the Convention on Biological Diversity. Professor Jefferson's Paper is an exhaustive review of the terminator genes, referred to in his paper as "genetic use restriction technologies" and abbreviated to GURTs.

> A patent only confers a *negative* right on its proprietor to prevent others from using the protected invention for a limited period. The right to positively use or not the invention by the patent holder is, hence, not addressed by the patent law which is primarily an instrument for promoting research by ensuring the possibility of excluding imitation by third parties.
>
> Hence, if the GURT patent were to be found inviable or invalid on any grounds, the effect of non-protection would be that the relevant method would remain or be put in the public domain. The absence of protection would not automatically lead to stop the eventual adoption and diffusion of the GURT technology; on the contrary, such an absence may foster its dissemination.

In less elegant words, what has been invented cannot be uninvented. Even if a patent is cancelled on the grounds that the invention is immoral, the inventor is still able to use the invention. Indeed, everyone is free to use it. Patent law is not the route to regulating the use of biotechnology.

The Way Forward

The case for strengthening intellectual property rights in the developing world is, I believe, overwhelming, and the need to strengthen these rights is urgent. Countries should evaluate their positions without delay, and must be urged and helped to implement TRIPs—or better, to improve upon TRIPs—as soon as possible. That said, the modification of intellectual property laws requires money and skilled advisors, both of which may be in short supply in developing countries.

Funding and other assistance is available. The World Intellectual Property Organisation (WIPO) has an agreement with the World Trade Organisation (WTO) to provide assistance to developing countries to meet their TRIPs commitments, to provide technical assistance in drafting, and to train staff and provide software. However, the limited resources of WIPO and WTO remain a constraint and could mean that some countries are unable to meet their commitments, even by the dead-line in the year 2005.

"The case for strengthening intellectual property rights in the developing world is, I believe, overwhelming."

The World Bank could certainly help by providing resources to the WTO or directly to the developing countries for this purpose. Such funding would not only increase the number of countries meeting their TRIPs obligations by the dead-line or even earlier, and it would also promote open and constructive discussion of TRIPs in the next WTO round which will begin early in the new century.

There is a further and additional approach which could save both time and resources. There is little logic in many different nations, each having similar standards and economic goals, to examine and grant patents for the same invention. The most efficient and economic approach is a regional organisation, such as the African Regional Intellectual Property Office (ARIPO), the African Intellectual Property Organisation (OAPI) or the European Patent Convention, which centralise the examination and granting of patents for all member countries into one office.

Stronger Intellectual Property Laws Help Everyone

Enforceable and strong intellectual property rights are essential for encouraging the transfer of the latest technologies to the developing world, and for stimulating research in these same new technologies. They are vital for the modernisation of crop production in the developing world.

Weak intellectual property laws and the inability to enforce intellectual property rights will limit the access of developing countries to the new technology which is so important for the development of their agriculture and the saving of valuable environmental resources. Weak laws will inhibit much-needed inward investment.

Each country must evaluate its own intellectual property laws and needs carefully, but I would urge all developing countries to strengthen these rights as soon as possible, and for the World Bank and other funding agencies to assist them in this endeavour. I am confident that the benefits—the access to the new technologies and inward investment—will follow.

Patents on Human Genes Are Ethical and Necessary

by M. Andrea Ryan

About the author: *M. Andrea Ryan is vice president of the Warner-Lambert Company, a drug company, and president of the American Intellectual Property Law Association (AIPLA).*

Mr. Chairman:

The American Intellectual Property Law Association (AIPLA) congratulates you and the Subcommittee for holding this oversight hearing on the relationship between the human genome and the United States patent system. We are particularly pleased to have the opportunity to offer our thoughts on this very timely and important subject.

The AIPLA is a national bar association whose more than 10,000 members are primarily lawyers in private and corporate practice, in government service, and in the academic community. The AIPLA represents a wide and diverse spectrum of individuals, companies, and institutions involved directly or indirectly in the practice of patent, trademark, copyright and unfair competition law, as well as other fields of law affecting intellectual property. Our members represent both owners and users of intellectual property.

A Successful History

AIPLA members practicing in the area of biotechnology are acutely aware of the important public policy issues that the Subcommittee is examining. AIPLA believes that these policy issues can be fully addressed by a straightforward application of existing principles of basic patent law.

The U.S. Constitution gives Congress the power to enact laws to protect the rights of inventors. These rights are granted and exist under the Constitution "to promote the progress of science and the useful arts." The foresight of the drafters of the Constitution in setting out a "patent clause" not only led in 1790 to the establishment of the U.S. patent system, but the action of the first of our

Excerpted from M. Andrea Ryan's testimony before the Subcommittee on Courts and Intellectual Property of the Committee on the Judiciary, at an oversight hearing on "Gene Patents and Other Genomic Inventions." July 13, 2000.

106 Congresses stands even today as a model throughout the world for promoting innovation. Over the years the U.S. patent system has been improved by the passage of more effective patent laws, such as the American Inventors Protection Act (AIPA), which passed in 1999. The patent system has also been invigorated by the faithful manner in which these laws have been implemented by the courts, notably the Court of Appeals for the Federal Circuit—a new appellate court created in 1983. For the most part, the U.S. patent system in its present form works effectively as an incentive to innovators. AIPLA believes that it will continue to work

> *"Patent protection should be made available to anything 'under the sun that is made by man.'"*

well even as technology changes and becomes more complex. Finally, we believe that where changes in the patent laws are needed, Congress will make them as it has many times over the past two decades. The Patent Law Amendments Act of 1984, the Drug Price Competition and Patent Term Restoration Act of 1984, the Patent Process Amendments Act of 1988, the Uruguay Round Agreements Act, and the American Inventors Protection Act of 1999 are all examples of innovative and responsive changes that have worked to improve the operation and effectiveness of the U.S. patent system.

Patenting Living Things

The issues surrounding patent protection relating to newly discovered genes and the often novel proteins that are products of the expression of these genes raises public policy and substantive law issues of patent law that are in many respects even more complex than the issue addressed by the Supreme Court in 1980 in *Diamond v. Chakrabarty*. In that case, the Supreme Court decided whether the fact that an invention was living should exclude the invention from the definition of patentable subject matter. The Supreme Court answered that question in the negative, by a five-to-four margin, and stated in effect that patent protection should be made available to anything "under the sun that is made by man". Those words and that court decision were instrumental in launching the modern biotechnology industry—and establishing the preeminent role of the United States as a leader in that industry.

The biotechnology patents that issued in the years that followed that important decision and the rapidly evolving technology brought us to the issue you are examining today. AIPLA believes that *Chakrabarty* was correctly decided by the Supreme Court. The Court's decision was firmly grounded in the legislative history of the 1952 Patent Act. The Congressional intent that "everything under the sun made by man" should be patentable was long applied to chemical substances. It is, therefore, inconceivable the Congress would permit the patenting of a genetically modified microbe that makes a life-saving drug, such as insulin, but not allow the person discovering the insulin gene to obtain a patent claiming the gene itself. Merely because there are clear and compelling policy justification for allowing patents related to genes to be patented does not, however, answer the most difficult questions: how broad should the protection afforded by such patents be, what work must be completed to make a gene-related invention ready for patenting, and how should patents of this type impact on research directed to understanding the gene and the full complexity of its biological role and functioning? Patent law has traditionally treated all biological materials—even genes—as chemicals, or "compositions of matter"—a traditional category of patent-eligible "invention." Patents, however, do not extent to products of nature, as such. Thus, naturally occurring biological substances have traditionally been patented once they have been isolated and identified as useful for a specific purpose or a specific function. At that point, a naturally occurring biological material, such as a gene, a hormone, an enzyme or the like can only be patented in the isolated or purified form that does not exist in nature. According to the Supreme Court ruling in *Chakrabarty* and established patent law, any product of nature is patentable if it is transformed in some way by man and it is also new, useful, and non-obvious.

The isolated or purified biological product cannot be validly patented unless the patent application that contains a claim to the product provides an adequate written description of the invention. Further, the disclosure in the patent application must enable persons skilled in biotechnology to make and use the claimed product. Some

> *"Some accounts in the popular press reflect a confusion concerning basic patent law principles."*

real-world utility for the claimed product must also be set out in the patent application—in some presently available form. Thus, for several decades, the patent law issue has not been whether an isolated or purified product obtained from nature, such as a gene-based invention, is eligible for patenting or is adequately disclosed in a patent application, but, rather, what is the proper form and scope of the application and claims for the patent to be granted?

Patenting Human Genes

A great deal has been written recently both in the popular press and in respected scientific journals on the topic of granting patents to inventions that relate to human genes and—most particularly—gene fragments. Some accounts in the popular press reflect a confusion concerning basic patent law principles and have generated much misinformation on this issue. Even worse, some accounts of the workings of the patent system have been erroneous and regrettably inflammatory.

In order to understand the patent issues raised, most recently by the publicity surrounding the human genome project and related subject matter, it is important to understand the basic science that leads to the inventions for which patent protection is being sought. One key to understanding biotechnology is in understanding the terms and definitions.

Any complex living being is made up of trillions of cells and inside every cell is a nucleus which contains a set of chromosomes. The information contained in all of the chromosomes in a cell is the genome of that being. The genome is the complete set of information for building and maintaining life of every organism. The genome contains the master blueprint for creating all cellular structure and activities for the living organism. The chromosomes which contain the genome are tightly coiled threads of DNA (deoxyribonucleic acid) and associated protein molecules. Each DNA molecule consists of two twisted strands of nucleotides and each strand contains many genes. A gene is a specific sequence of nucleotide bases. Genes encode for protein and express (produce) that protein. Proteins are the building blocks of nature and combine to make up cells which grow to create life. A gene per se [in itself] is not life.

Interspersed with the genes which carry the essential protein coding information are intron sequences which have no apparent coding function and are sometimes referred to as "Junk DNA". Genes make up only a small percentage of the genome. It is this small percent of the genome that is the focus of so much attention and generally raises the questions about granting patents. The genome is a map of the entire area, but what is most interesting to scientists and the patent attorneys who work with them is not the genome, but the genes, portion of genes, and the sequences of nucleotides that a gene is made up of, including SNPs (single nucleotide polymorphisms) which are variations which occur in the DNA sequence of the gene and ESTs (expression sequence tags). Scientists believe that these tools will help them to identify the multiple genes associated with complex diseases and to design better and more specific drugs and treatments for these diseases. Much of the discussion about the use of genes and related inventions is still speculation. Genes and the other related gene technology may or may not turn out to be good targets for drug design. It may be possible to design drugs from the genetic information, but that will take years to determine. Not only must scientists determine what each gene does, but also precisely which proteins each gene produces. In order to design a drug, they will also need to know the structure and the function of the proteins which is expressed (produced) by the gene. This has been determined for some proteins, but in addition to knowing the protein structure, the actual folding of the protein also appears to be important in designing drugs to cure specific diseases. Determining the way a protein folds is apparently a difficult job.

Much Work to Be Done

Steven Holyman of Millennium Pharmaceuticals focused the discussion on the right place when he recently said "the race is in assigning to genes and to variations in genes a role in disease initiation and progress and drug response".

Nobel Prize winner, David Baltimore apparently agrees. In a recent article in the *New York Times* he recognized that "the sequencing of the genome is a landmark of progress in specifying information, decoding it into its many coded meanings, and learning how it goes wrong in disease. While it is a moment worthy of the attention of every human, we should not mistake progress for a solution. There is yet much hard work to be done".

Applying basic patent law to these concepts means that in spite of the fact that patents are being applied for and granted now, there is still much more to be discovered and those discoveries should not only be patentable, but valuable. One of the touchstones of patentability is whether or not the invention is the solution to a problem and if so, how difficult was the problem to solve. Paraphrasing Dr. Baltimore, the genome is not the solution; years of work remain to be done by researchers to apply genomic and gene related inventions to curing diseases and designing drugs which also in themselves should be patentable.

The patent issues surrounding biotechnology and specifically genes and gene-related technology are less than 20 years old in 2000 and it will take time to sort out the application of the patent laws to this technology. As the United States Patent and Trademark Office ("the USPTO") works its way through the new applications and the Courts deal with challenges to already issued patents, more clarity will be found. Time will help—there is no immediate need to solve all of these issues. The patent law like the technology will evolve and grow to address the new issues. The Court of Appeals for the Federal Circuit is addressing and will continue to address the questions of which genes, SNPs, ESTs and other genetic material are patentable. The Courts and the USPTO need time to work through these issues and they have the basic tools that they need to do this.

Clarifying Patent Law

AIPLA approves of the USPTO's efforts to clarify and provide for consistency in the training of its examiners as to the manner in which the written description requirement and the utility requirement for a patent application is to be applied to the examination of patent applications. An adequate written description is fundamental to the proper functioning of the patent system. The full benefits of a patent cannot be realized if it does not contain a written description which discloses the "manner and process of making and using an invention in such full, clear, concise, and exact terms as to enable any person skilled in the art" to make and use the invention. This is particularly critical in the area of patents and patent applications involving genes, gene sequences, and related biotechnological inventions.

> *"One of the touchstones of patentability is whether or not the invention is the solution to a problem."*

We believe the Revised Written Description Guidelines and the Utility Guidelines as published by the Office have taken great steps forward in the complex area of the written description requirements for a biotechnology patent. The

AIPLA urges that patent examiners should be instructed in the Revised Written Description Guidelines to exercise vigilance and to make rejections of patent claims on written description grounds whenever there is a clear and reasonable basis for doing so. If the USPTO fails to exercise vigilance in the identification and rejection of written description defects, patents with invalid, overly broad claims could be issued, spawning expensive and time consuming litigation that could have been avoided. Similarly, where the ex parte [one-sided] appeal process from USPTO decisions results in a decision that is favorable to inventors on written description grounds, the clear validity of the resulting claims will similarly reduce litigation and provide useful guidance to examiners.

AIPLA strongly urges the USPTO to follow the decisional law of the 1990s that in certain respects has elevated the importance of the written description and utility requirements and use this guidance to reject claims in applications or invalidate claims in patents. AIPLA believes that it would be preferable for the law on the written description and utility requirements to be developed at an early stage through ex parte appeals from the USPTO rather than through later, more expensive post-issue litigation in the Federal Courts. This belief necessarily translates into a desire to see the USPTO rigorously apply the statutory written description and utility requirements as applied by the Federal Circuit. Moreover, AIPLA would urge the USPTO to identify appeals on these issues and expedite their disposition within the USPTO, to the extent consistent with law and regulation.

Preparing for Difficult Decisions

If anything in the patent system needs to be changed immediately to address the issues surrounding the patenting of genes, pieces of genes and in fact all of the burgeoning new science of biotechnology, it is increased support for the USPTO. The patent examiners are the individuals who make the initial and critical decisions as to when to grant or not to grant a U.S. patent. If the USPTO does not have the funds to continue to hire and train these examiners, the quality and quantity of patents granted in this important technology will be seriously impacted. If Congress continues to divert funds from the Office budget, the hiring, training, and retention of examiners—and ultimately the quality of biotechnology applications—will suffer. . . .

In addition, AIPLA believes that additional changes to the current patent law could also help to address some of the questions raised by the granting of patents in the area of genes and related new subject matter. AIPLA recognizes and commends the efforts of this Subcommittee for striving to achieve some of the changes which would be beneficial in this regard. H.R. 400, as reported by this Subcommittee in 1999, contained what promised to be a very helpful expansion of the existing reexamination system. It would have allowed members of the public limited participation in the reexamination process before the USPTO, including the ability to appeal and to participate in appeals in the

Office and before the Court of Appeals for the Federal Circuit. This would have provided a very cost effective means of challenging problematic patents granted in this area. Unfortunately, the inter partes reexamination procedures were drastically curtailed during the subsequent legislative deliberations that led to the AIPA. Another limitation . . . involves the exception to 18-month publication in the AIPA. Full publication of all pending applications would provide researchers and companies in the biotechnology field greater certainty regarding their freedom to pursue costly and expensive research in this field.

No Bar to Research

Questions have been raised regarding whether there are any unintended impacts of the existing patent laws on basic scientific research and on the freedom of doctors to use new gene related inventions in the treatment of patients. This latter concern was the subject of legislation in 1996 when Congress excluded from the definition of infringement surgical and medical procedures that a doctor might perform on a patient. While AIPLA opposed that amendment on the basis that it was based on only a single example of dubious real-world significance and was inconsistent with the obligations of TRIPs [the Agreement on Trade Related Aspects of Intellectual Property Rights, established in 1995], it is nonetheless the law and should remove this concern from the list of allegedly harmful consequences of granting patents in the area of gene and related inventions. Regarding the allegation that the patent laws may have unintended adverse impacts on basic scientific research, certainly the explosion of investment in the biotech field would not support such a conclusion. To start with, patents never "lock up" information or prevent the use of gene sequence information in any context. The patent system is designed to assure that information gets disclosed to the public rapidly. The information concerning an invention—what it is, how it can be made, what it is useful for—go immediately into the public domain, free for all to use. Thus, the patent system has led to the publication of massive quantities of information concerning genes and

> *"The patent system is designed to assure that information gets disclosed to the public rapidly."*

their function. Everyone has free and unfettered access to that information. While inventions can be patented, information cannot.

Second, the Supreme Court has long recognized that not all "uses" of a patented invention represent an infringement of the patent owner's rights. Although very limited, an "experimental use" exemption does exist. It has been developed by the courts to assure that a patented invention can be used to understand the basic function of the invention and develop alternatives to it. If only as a logical matter, the patent system can never promote progress in the useful arts if the grant of a patent locks out others from gaining a basic understanding of what is patented, how to design around it, and how to improve upon it. Absent an experimental use exception, patents could theoretically freeze, not promote progress in the useful arts and frustrate the development of improvement inventions that Congress has specifically authorized to be patented. Many commentators believe that this so-called "experimental use" or "research exemption" under current case law is sufficient to assure that all basic research activities can peacefully co-exist with the broad, exclusionary rights of the patent owner to stop unauthorized uses of a patented invention.

> *"The U.S. patent system is providing unprecedented hope for the nation's sick and infirm."*

In this regard, gene research and gene patents interplay no differently compared to research and patents in other technological fields. Similar concerns have been raised in many technological areas. As of mid-2000, we have not seen emerging technology fields blocked, locked down or frozen in place by a "pioneer" patent. One very likely reason that this has not occurred in other fields is the reality of pioneer innovators. Their seminal inventions often take years to bring to full fruition through wide adoption in the marketplace. This requires that the technology be developed quickly since, once a patent issues, the patent term winds down over the course of only a few years. For many seminal inventions, such as Stanford's Cohen-Boyer patents, this has meant providing licenses to the entire industry—hoping to spur development of implementing technology.

If an invention does, however, fall within the scope of an earlier valid dominating patent or was discovered as a result of using an earlier patented invention, the later inventor/patentee may need to obtain a license, if one is available, and to pay a royalty to the owner of the earlier patent—but this is no different than in any other technology field—including the other explosively growing fields, telecommunications, Internet, software, and semiconductor-based devices. The patent system has worked in the past and, given time and the reality of the marketplace, it should work in this field as well. In brief, one should be rather chary about designing solutions in search of a problem for our time-tested and venerable patent system. Should such a problem materialize in the future, it can be appropriately dealt with at that time.

A Spur to Invention

The U.S. patent system is providing unprecedented hope for the nation's sick and infirm while serving the biotechnology community. The USPTO has demonstrated that it is aware of the needs of everyone impacted by the patent system. It is seeking to improve its processing of gene and related patent applications. As indicated, however, the Office has a desperate need for all of the fee revenue it receives to keep pace with its ballooning workload in this complex field and improving the quality of its work. We are confidant that with your assistance, the issue of consistent and adequate PTO funding will be successfully addressed. We believe that the Office is targeting test cases to clarify the utility and written description questions that are outstanding [still unsettled], and that the experience and competence of the Court of Appeals for the Federal Circuit will aid immeasurably in providing any needed guidance. We will certainly be monitoring the developments in this very important and rapidly moving field and look forward to working with this Subcommittee to resolve real issues as they arise.

In summary, we urge the Congress to stay the current course in terms of the patent laws themselves. The patent system is working in this area as it has worked effectively elsewhere—to make information on new inventions promptly available to spur further innovation, to provide incentives for investments that will produce new businesses and new products, and—ultimately—to secure the blessing of accelerating innovation for ourselves and our posterity.

Glossary

allele   A particular form of a gene.

allergen   A substance capable of causing an allergic reaction, in which the body attempts to defend itself against the "foreign" material in ways that cause harm (producing sneezing, skin rashes, or more serious problems).

***Bacillus thuringiensis* (Bt)**   A bacterium that produces a substance poisonous to many pest insects; the gene that makes this substance has been engineered into some crop plants.

biodiversity   The variety or diversity of kinds of living things in a particular area; greater diversity is thought to produce a healthier environment or ecosystem.

biotechnology   The use or alteration of living things in industrial processes; often used as a synonym for **genetic engineering**.

Brave New World   A science fiction novel written by Aldous Huxley, published in 1932. It portrays a future dictatorship that, among other things, has made reproduction a laboratory process and controls the genetics of all children.

chromosome   One of a group of threadlike bodies in the nucleus (central body) of a cell that contains a living thing's hereditary information (genes).

clone   An exact genetic copy of a cell or a living thing.

DNA   Deoxyribonucleic acid, the chemical of which genes are made.

Dolly   A sheep cloned from an udder cell of an ewe (female sheep) in Scotland in 1997; she was the first mammal to be cloned from a mature body cell in an adult animal.

embryo   A living thing in a very early stage of development before birth; in humans, an unborn child during the first two months of development is considered to be an embryo.

embryo selection   A process in which several eggs are fertilized (united with sperm) in the laboratory, the resulting embryos are examined to see whether they contain disease-causing genes, and only healthy embryos are implanted in a uterus and allowed to develop.

entomologist   A scientist who studies insects.

Environmental Protection Agency (EPA)   One of three federal government agencies in the United States that regulates genetically engineered crops; it regulates any crops that contain pesticides, such as Bt.

eugenics   A belief, popular in the late nineteenth and early twentieth centuries, that the human race could be improved by encouraging people with desirable characteristics to reproduce and discouraging or forcibly preventing those with characteristics considered undesirable from reproducing.

fetus   A living thing in the later stages of development before birth; in humans, an unborn child is considered to be a fetus after the first two months of development (before that, it is an **embryo**).

Food and Drug Administration (FDA)   One of three federal government agencies in the United States that regulates genetically altered products. The FDA regulates products that will be used in food or medicine and tests of treatments for disease that involve altering genes (gene therapy).

"Frankenfoods"   Derogatory term for foods containing genetically engineered material, used especially in Europe, referring to the monster in *Frankenstein*.

Frankenstein   A novel by Mary Shelley, published in 1818, in which a scientist, Victor Frankenstein, creates a monster from parts of dead bodies and brings it to life.

gene   A unit of inherited material that carries coded instructions for making one protein or doing one other job in the body, such as controlling another gene.

gene therapy   Treating or preventing disease by altering genes.

genetic code   The basic code by which the sequence, or order, of small molecules called bases within a molecule of DNA carries inherited information that determines the characteristics of a living thing.

genetic engineering   Alteration of the genes of a living thing in a laboratory, for instance by inserting a gene from a different kind of living thing.

genome   A living thing's complete collection of genes.

genotype   The characteristics of a living thing as determined by its genes.

germline genetic engineering   Changing the genes in sperm or egg cells, causing the changes to be passed on to future generations of living things; compare **somatic cell genetic engineering.**

GM foods   Foods that have been, or contain material that has been, genetically modified (genetically engineered).

Green Revolution   New crop varieties and technologies introduced during the 1960s in the hope of increasing crop yields and, therefore, food supply; this approach was later criticized because it required heavy use of pesticides and fertilizers, which raised costs and damaged the environment.

greens   A slang term for environmentalists, especially those who are politically active.

herbicide   A chemical that kills plants.

intellectual property   Something produced by the mind, which can be protected by a patent, copyright, trademark or similar legal procedure.

in vitro fertilization (IVF)   Fertilization (joining of a sperm and an egg) that takes place in a laboratory rather than within a living body.

mad cow disease   A brain-destroying disease caused by eating nerve tissue from infected animals; an epidemic of this disease occurred among cattle in Britain in the 1990s, and a small number of humans are believed to have caught the disease by eating infected meat.

monoculture   A farming practice in which large areas are planted with the same crop.

moratorium   Banning of an activity for a limited period of time.

National Institutes of Health (NIH)   A group of large research organizations in Bethesda, Maryland, sponsored by the government of the United States.

NGOs   Nongovernment organizations, such as nonprofit groups.

organic food   Food grown with only animal or vegetable fertilizers or pesticides.

oversight   Monitoring and regulation.

patent   The exclusive right to make, sell, or profit from an invention or process for a certain length of time.

pharmaceutical   Medicine or drug.

protein   One of a large group of chemicals that do most of the work of the body; proteins are made in cells according to instructions carried in genes.

protocol   A planned series of procedures.

recombinant bovine growth hormone (RBGH)   A growth hormone of cattle, made artificially by bacteria into which the cattle gene for this hormone has been inserted, given to dairy cattle to make them produce more milk.

recombinant DNA   DNA that contains genes from more than one kind (species) of living thing.

reductionism   A philosophy that reduces complex objects or processes to their simplest parts or terms, for instance considering living things strictly in terms of physical and chemical reactions.

slippery slope   A philosophical idea referring to a situation in which a first, ethically unobjectionable action leads unavoidably to other actions that are more and more ethically questionable.

somatic cell gene therapy (genetic engineering)   Alteration of genes in certain body cells, such as blood cells, but not in the sex cells (sperm and eggs), which carry the genes that will be passed on to future generations; the gene changes therefore affect only a particular individual, not his or her offspring. Compare **germline genetic engineering**.

staple   A basic, everyday food.

subsistence farming   Farming done only or primarily to feed one's family, rather than to produce crops to sell.

sustainable farming   Farming that sustains a population without depleting natural resources or damaging the environment.

transgenic   Containing genes from more than one kind (species) of living thing.

U.S. Department of Agriculture (USDA)   One of three agencies of the United States government that regulates genetically engineered products. The USDA, sometimes together with the EPA and the FDA, regulates genetically altered plants and animals used in agriculture (farming).

vector   Something that carries a gene into a different living thing; viruses are the most common vectors.

Organizations to Contact

The editors have compiled the following list of organizations concerned with the issues debated in this book. The descriptions are derived from materials provided by the organizations themselves. All have publications or information available for interested readers. The list was compiled on the date of publication of the present volume: names, addresses, phone and fax numbers, and e-mail and Internet addresses may change. Be aware that many organizations take several weeks or longer to respond to inquiries, so allow as much time as possible.

AgBioWorld
website: www.agbioworld.org

AgBioWorld seeks to educate scientists, the media, and the public about the benefits of agricultural biotechnology. Papers available on its website include "Viewpoints on Biotechnology" and "Responses to GM Food Myths."

Agricultural Groups Concerned About Resources and the Environment (AGCare)
e-mail: agcare@agcare.org • website: www.agcare.org

This group represents the crop producers of Ontario, Canada. Its website contains numerous papers on agricultural biotechnology, including some aimed at answering consumers' questions, such as "Introduction to Agri-Food Biotechnology" and "Scientists Urged to Defend Genetically Engineered Food." It generally favors biotechnology.

American Crop Protection Association (ACPA)
1156 Fifteenth St. NW, Suite 400, Washington, DC 20005
(202) 296-1585 • fax: (202) 463-0474
website: www.acpa.org

The ACPA promotes the environmentally sound use of crop protection products, including bioengineered plants containing Bt and other pesticide genes, for the economical production of safe, high-quality, abundant food and other crops. It represents the pesticide industry. Its website includes general information about plant biotechnology and papers and news releases describing the benefits and safety of genetically engineered crops, such as "What Experts Say About Biotechnology."

American Society of Law, Medicine and Ethics
765 Commonwealth Ave., Suite 1634, Boston, MA 02215
(617) 262-4990 • fax: (617) 437-7596
e-mail: info@aslme.org • website: www.aslme.org

This group acts as a forum for discussion of issues including the ethics of genetic engineering. Its material is aimed primarily at professionals in the fields of health care and law. It publishes two quarterly journals, *Journal of Law, Medicine, and Ethics* and *American Journal of Law and Medicine*. Its website includes papers on deciphering and engineering the human genome, including "Genetics and Society: Impact of New Technologies on Law, Medicine, and Policy."

Biotechnology Industry Organization (BIO)
1625 K St. NW, Suite 1100, Washington, DC 20006
(202) 857-0244
website: www.bio.org

The Biotechnology Industry Organization represents biotechnology companies, academic institutions, state biotechnology centers, and related organizations that support the use of biotechnology in agriculture, health care, and other fields. BIO works to educate the public about biotechnology and responds to concerns about the safety and ethics of genetic engineering and related technologies. Its website includes an introductory guide to biotechnology as well as links to other biotechnology websites and press releases and position papers on bioethics, food and agriculture, intellectual property issues, and similar topics.

Center for Bioethics and Human Dignity (CBHD)
2065 Half Day Rd., Bannockburn, IL 60015
(847) 317-8180 • fax: (847) 317-8153
e-mail: cbhd@cbhd.org • website: www.bioethix.org

CBHD is an international education center whose purpose is to bring Christian perspectives to bear on contemporary bioethical challenges facing society. It opposes the alteration of human genes. It publishes the newsletter *Dignity* as well booklets, videos, and other materials. Its website contains articles (for example, "Genetic Intervention: The Ethical Challenges Ahead") and public statements, issue overviews, bibliographies, links, and publications for sale.

Center for Food Safety and Applied Nutrition
200 C St. SW, Washington, DC 20204
(888) 463-6332
website: www.cfsan.fda.gov

The Center for Food Safety and Applied Nutrition is the part of the federal government's Food and Drug Administration (FDA) that regulates genetically engineered food crops. The center's website includes a collection of papers on biotechnology and the FDA's regulation of the biotechnology industry, including "Draft Guidance for Industry: Voluntary Labeling Indicating Whether Foods Have or Have Not Been Developed Using Bioengineering," posted January 2001, and "FDA and CDC Reports to EPA on Starlink Corn," posted June 2001.

Council for Biotechnology Information
website: www.whybiotech.com

The Council for Biotechnology Information was founded by leading biotechnology companies to provide information about the benefits of this technology. Its website includes numerous papers supporting this point of view.

Council for Responsible Genetics (CRG)
5 Upland Rd., Suite 3, Cambridge, MA 02140
(617) 868-0870 • fax: (617) 491-5344
e-mail: crg@gene-watch.org • website: www.gene-watch.org

The Council for Responsible Genetics is a national nonprofit organization of scientists and others devoted to encouraging public debate about the social, ethical, and environmental implications of new genetic technologies. It works to provide members of the public with clear and understandable information on genetic innovations so that they can participate in decision making about genetic technology and its implementation. Material on CRG's website includes a petition titled "No Patents on Life," a position paper on manipulation of the human germline, and news alerts.

Foundation on Economic Trends (FET)
1660 L St. NW, Suite 216, Washington, DC 20036
(202) 466-2823 • fax: (202) 429-9602
e-mail: office@biotechcentury.org • website: www.biotechcentury.org

Founded by biotechnology critic and author Jeremy Rifkin, the foundation is a nonprofit organization whose mission is to examine emerging trends in science and technology and their impacts on the environment, the economy, culture, and society. FET works to educate the public about topics such as gene patenting, commercial eugenics, genetic discrimination, and genetically altered food. It has proposed a moratorium on somatic (individual) gene therapy using viruses to transmit genes into human cells and favors labeling of genetically altered food. Information on these positions is available on the organization's website.

Future Generations
e-mail: vancourt@eugenics.net • website: www.eugenics.net

This group, headed by Marian Van Court, strives to leave a legacy of good health, high intelligence, and noble character to future generations by humanitarian eugenics, or judicious altering of the human genome. The organization's website includes a number of articles explaining and defending its point of view.

Human Cloning Foundation
PMB 143, 1100 Hammond Dr., Suite 410A, Atlanta, GA 30328
e-mail: RWicker@gateway.net • website: www.humancloning.org

The foundation is a nonprofit organization that promotes education about human cloning and gene alteration and emphasizes the positive aspects of these technologies. Its website contains numerous articles and fact sheets supporting human cloning.

Institute for Food and Development Policy (Food First)
398 60th St., Oakland, CA 94608
(510) 654-4400 • fax: (510) 654-4551
e-mail: foodfirst@foodfirst.org • website: www.foodfirst.org

The Institute for Food and Development Policy, better known as Food First, is a member-supported, nonprofit think tank and education-for-action center working to highlight root causes and value-based solutions to hunger and poverty around the world. It produces books, reports, articles, films, workshops, and other educational material and action plans for the public, policy makers, activists, students, and the media. Papers available on its website include "Genetic Engineering of Food Crops for the Third World" and "Critiquing Biotechnology and Industrial Agriculture."

International Food Information Council Foundation (IFIC)
1100 Connecticut Ave. NW, Suite 430, Washington, DC 20036
(202) 296-6540 • fax: (202) 296-6547
e-mail: foodinfo@ific.health.org • website: www.ific.org

IFIC aims to bridge the gap between science and communications by collecting and distributing scientific information on food safety, nutrition, and health to opinion leaders and consumers. Its website includes newsbriefs such as "Agricultural Biotechnology Provides a New Rice for Africa." Single copies of publications, such as "Food Biotechnology: Enhancing Our Food Supply," may be ordered free online.

International Food Policy Research Institute (IFPRI)
2033 K St. NW, Washington, DC 20006-1002
(202) 862-5600 • fax: (202) 467-4439
e-mail: ifpri@cgiar.org • website: www.ifpri.cgiar.org

IFPRI is the United States center of the Consultative Group on International Agricultural Research (CGIAR). Its mission is to identify and analyze policies for sustainably meeting the food needs of the developing world. It publishes a twice-yearly newsletter, *Globalization,* and booklets such as "World Food Prospects," some of which can be ordered free online.

National Bioethics Advisory Commission (NBAC)
6705 Rockledge Dr., Suite 700, Rockville, MD 20892-7979
(301) 402-4242 • fax: (301) 480-6900
website: www.bioethics.gov

NBAC is a federal agency that sets guidelines governing the ethical conduct of research. It works to protect the rights and welfare of human research subjects and governs the management and use of genetic information. Its published reports include *Cloning Human Beings* and *Ethical Issues in Human Stem Cell Research.*

National Human Genome Research Institute (NHGRI)
9000 Rockville Pike, Bethesda, MD 20892
(301) 402-0911 • fax: (301) 402-0837
website: www.nhgri.nih.gov

Sponsored by the National Institutes of Health, the federal government's primary agency for the support of biomedical research, NHGRI heads the Human Genome Project, the federally funded effort to map all human genes. Information about the Human Genome Project, including its ethical, legal, and social implications, is available at NHGRI's website.

Novartis Foundation for Sustainable Development
e-mail: novartis.foundation@group.novartis.com
website: www.foundation.novartis.com

This organization, sponsored by the international life science conglomerate Novartis, is engaged in programs in developing countries that contribute to improving the quality of life for the poorest people. Articles available on its website include "Ethical and Ecological Aspects of Industrial Property Rights in the Context of Genetic Engineering and Biotechnology."

Organic Consumers Association
6101 Cliff Estate Rd., Little Marais, MN 55614
(218) 226-4164 • fax: (218) 226-4157
website: www.purefood.org

The Organic Consumers Association is a national grassroots organization dealing with issues of food safety, industrial agriculture, genetic engineering, and environmental sustainability. It opposes genetic engineering of foods. The organization publishes a newsletter, *BioDemocracy News,* every six weeks. Its website includes papers on genetically engineered food, such as "Hazards of GE Foods and Crops and Guidelines for Grassroots Action" and "What's Wrong with Genetic Engineering."

Rural Advancement Foundation International (RAFI)
478 River Ave., Suite 200, Winnipeg MB R3L OC8 Canada
(204) 453-5259 • fax: (204) 284-7871
e-mail: rafi@rafi.org • website: www.rafi.org

RAFI is an international, nongovernmental organization dedicated to the conservation and sustainable improvement of agricultural biodiversity and to the socially responsible development of technologies useful to rural societies. It opposes agricultural biotechnology, especially as managed by large companies, and patenting of genetic material. It publishes a communique four to six times a year as well as occasional papers and other publications, many of which are available online. An example is "In Search of Higher Ground: The Intellectual Property Challenge to Public Agricultural Research and Human Rights."

U.S. Department of Agriculture (USDA)
Animal and Plant Health Inspection Service (APHIS)
14th and Independence Ave. SW, Washington, DC 20250
e-mail: john.t.turner@usda.gov • website: www.aphis.usda.gov/biotechnology

The USDA is one of three federal agencies, along with the Environmental Protection Agency (EPA) and the Food and Drug Administration (FDA), primarily responsible for regulating biotechnology in the United States. The USDA's Animal and Plant Health Inspection Service (APHIS) conducts research on the safety of genetically engineered organisms, helps form government policy on agricultural biotechnology, and provides information to the public about these technologies. The APHIS website includes policy statements on biotechnology; a description of the roles of the USDA, FDA, and EPA in regulating agricultural biotechnology; and research reports, including "Impacts of Adopting Genetically Engineered Crops in the United States."

Additional Internet Resources

The following websites contain a wealth of information for students and others interested in learning more about genetic engineering and ethical issues involving this technology.

Access Excellence
website: www.accessexcellence.org

This site is aimed at teachers and students and contains news about biotechnology, a history of biotechnology (the Biotech Chronicles), educational activities, and links to many other sites related to biotechnology and genetics.

Bioethicsline
website: http://igm.nlm.nih.gov

Sponsored by the National Library of Medicine (NLM), part of the National Institutes of Health, Bioethicsline is an online medical database. It offers annotated bibliographies on the ethics of technologies such as gene therapy and human cloning. It is accessed through NLM's Internet Grateful Med website.

Biotechnology Information Resource Site
website: www.nalusda.gov/bic

This site is sponsored by the National Agricultural Library of the U.S. Department of Agriculture. It includes links to numerous papers and organizations related to agricultural biotechnology, such as "Genetically Modified Crops and Foods" by the American Medical Association's Council on Scientific Affairs.

National Biotechnology Information Facility (NBIF)
website: http://nbif.org

This website contains a variety of games, activities, and other materials that educate students about biotechnology and an Internet Resources section that includes more than 3,300 annotated links to biotechnology-related sites.

Bibliography

Books

William R. Clark	*The New Healers*. New York: Oxford University Press, 1997.
Eric S. Grace	*Biotech Unzipped: Promises and Realities*. Washington, DC: Joseph Henry Press, 1997.
Ruth Hubbard and Elijah Wald	*Exploding the Gene Myth*. Boston: Beacon Press, 1997.
Leon R. Kass and James Q. Wilson	*The Ethics of Human Cloning*. Washington, DC: AEI Press, 1998.
Glenn McGee	*The Perfect Baby: A Pragmatic Approach to Genetics*. Lanham, MD: Rowman & Littlefield, 1997.
Gregory E. Pence	*Who's Afraid of Human Cloning?* Lanham, MD: Rowman & Littlefield, 1998.
M. L. Rantala and Arthur J. Milgram, eds.	*Cloning: For and Against*. Chicago: Open Court, 1999.
Jeremy Rifkin	*The Biotech Century*. New York: Jeremy P. Tarcher, 1998.
Thomas A. Shannon	*Genetic Engineering: A Documentary History*. Westport, CT: Greenwood Publishing Group, 1999.
Thomas A. Shannon	*Made in Whose Image: Genetic Engineering and Christian Ethics*. Amherst, NY: Humanity Books, 1999.
Lee M. Silver	*Remaking Eden: How Genetic Engineering and Cloning Will Transform the American Family*. New York: Avon, 1998.
Gregory Stock and John Campbell, eds.	*Engineering the Human Germline: An Exploration of the Science and Ethics of Altering the Genes We Pass to Our Children*. New York: Oxford University Press, 2000.
Martin Teitel and Kimberly A. Wilson	*Genetically Engineered Food: Changing the Nature of Nature*. Rochester, VT: Park Street Press, 1999.
James D. Torr, ed.	*Opposing Viewpoints: Genetic Engineering*. San Diego, CA: Greenhaven Press, 2001.
Jon Turney	*Frankenstein's Footsteps: Science, Genetics and Popular Culture*. New Haven, CT: Yale University Press, 1998.

Casey Walker, ed.   *Made Not Born: The Troubling World of Biotechnology.* San Francisco: Sierra Club Books, 2000.

Lisa Yount   *At Issue: The Ethics of Genetic Engineering.* San Diego, CA: Greenhaven Press, 2001.

Lisa Yount   *Library in a Book: Biotechnology and Genetic Engineering.* New York: Facts On File, 2000.

Periodicals

W. French Anderson   "A Cure That May Cost Us Ourselves," *Newsweek*, January 1, 2000.

W. French Anderson   "Are Bio-Foods Safe?" *Business Week,* December 20, 1999.

Ronald Bailey   "Petri Dish Politics," *Reason*, December 1999. Available from The Reason Foundation, 3415 S. Sepulveda Blvd., Suite 400, Los Angeles, CA 90034.

Ronald Bailey   "The Twin Paradox," *Reason*, May 1997. Available from The Reason Foundation, 3415 S. Sepulveda Blvd., Suite 400, Los Angeles, CA 90034.

Paul R. Billings, Ruth Hubbard, and Stuart A. Newman   "Human Germline Gene Modification: A Dissent," *Lancet*, May 29, 1999.

Martin Bobrow and Sandy Thomas   "Patents in a Genetic Age," *Nature*, February 15, 2001.

Kathryn S. Brown   "Food with Attitude," *Discover*, March 2000.

Kevin Clarke   "Designer Babies, Anyone?" *National Catholic Reporter,* October 22, 1999. Available from *National Catholic Reporter*, 115 East Armour Blvd., Kansas City, MO 64111.

Kevin Clarke   "Unnatural Selection," *U.S. Catholic*, January 2000. Available from *U.S. Catholic*, 205 West Monroe St., Chicago, IL 60606.

Mike Fillon   "Grains of Hope," *Time,* July 31, 2000.

Mike Fillon   "Natural Born Factories," *Popular Mechanics*, March 1998.

Thomas Hayden   "Monkeying with Nature," *U.S. News & World Report*, January 22, 2001.

Mae-Wan Ho   "The Unholy Alliance," *The Ecologist*, July–August 1997.

Marguerite Lamb et al.   "Brave New Food," *Mother Earth News*, April 2000. Available from *Mother Earth News*, P.O. Box 56302, Boulder, CO 80322-6302.

Mary Midgley   "Biotechnology and Monstrosity: Why We Should Pay Attention to the 'Yuk Factor,'" *The Hastings Center Report*, September 2000.

Helena Paul   "Moral Bankruptcy: Adoption of the EU Life Patents Directive," *The Ecologist*, July–August 1998.

Bibliography

Virginia I. Postrel — "Fatalist Attraction: The Dubious Case Against Fooling Mother Nature," *Reason*, July 1997.

Susan Reed — "My Sister, My Clone," *Time*, February 19, 2001.

Jeremy Rifkin — "The Ultimate Therapy," *Tikkun*, May/June 1998. Available from *Tikkun*, 2107 Van Ness Ave., Suite 302, San Francisco, CA 94109.

Albert Rosenfeld — "New Breeds Down on the Pharm," *Smithsonian*, July 1998.

Ismael Sarageldin — "Biotechnology and Food Security in the 21st Century," *Science*, July 16, 1999.

Pamela Schaeffer — "Revolution in Biology Drives Revolution in Theology, Ethics and Law," *National Catholic Reporter*, October 22, 1999.

Vandana Shiva — "The Plunder of Nature," *Synthesis/Regeneration*, Spring 1999. Available from WD Press, Box 24115, St. Louis, MO 63130.

Lee Silver, Jeremy Rifkin, and Barbara Katz Rothman — "Biotechnology: A New Frontier of Corporate Control," *Tikkun*, July–August 1998. Available from *Tikkun*, 2107 Van Ness Ave., Suite 302, San Francisco, CA 94109.

Ricarda Steinbrecher — "What Is Wrong with Nature?" *Synthesis/Regeneration*, Winter 1999.

Mark Strauss — "When Malthus Meets Mendel," *Foreign Policy*, Summer 2000.

Laura Tangley — "Engineering the Harvest," *U.S. News & World Report*, March 13, 2000.

Laura Tangley — "Of Genes, Grain, and Grocers," *U.S. News & World Report*, April 10, 2000.

Larry Thompson — "Are Bioengineered Foods Safe?" *FDA Consumer*, January 2000.

Paul B. Thompson — "Food Biotechnology's Challenge to Cultural Integrity and Individual Consent," *Hastings Center Report*, July/August, 1997. Available from The Hastings Center, Garrison, NY 10524-5555.

Colin Tudge — "When It's Right to Be a Luddite," *New Statesman*, April 24, 2000.

Ian Wilmut — "Dolly's False Legacy," *Time*, January 11, 1999.

Roger Wrubel — "Biotechnology: Right or Wrong?" *Bioscience*, March 1998.

Index

Index